SpringerBriefs in Electrical and Computer Engineering

For further volumes:
http://www.springer.com/series/10059

Moamar Sayed-Mouchaweh

Discrete Event Systems

Diagnosis and Diagnosability

 Springer

Moamar Sayed-Mouchaweh
Départment Informatique et Automatique
Ecole des Mines de Douai
Douai cedex
France

ISSN 2191-8112   ISSN 2191-8120 (electronic)
ISBN 978-1-4614-0030-1   ISBN 978-1-4614-0031-8 (eBook)
DOI 10.1007/978-1-4614-0031-8
Springer New York Heidelberg Dordrecht London

Library of Congress Control Number: 2014932268

Printed on acid-free paper

Springer is part of Springer Science+Business Media (www.springer.com)

Preface

Discrete Event Systems: Diagnosis and Diagnosability addresses the problem of fault diagnosis of discrete event systems (DES). This book provides the basic techniques and approaches necessary for the design of an efficient fault diagnosis system for a wide range of modern engineering applications. The different techniques and approaches are classified according to several criteria such as: modeling tools (Automata, Petri nets) that is used to construct the model; the information (qualitative based on events occurrences and/or states outputs, quantitative based on signal processing and data analysis) that is needed to analyze and achieve the diagnosis; the decision structure (centralized, decentralized, distributed) that is required to achieve the diagnosis; The goal of this classification is to select the efficient method to achieve the fault diagnosis according to the application constraints. Then the book focuses on the centralized and decentralized event based diagnosis approaches using formal language and automata as modeling tool. This book includes illustrated examples of the presented methods and techniques as well as a discussion on the application of these methods on several real-world problems. This book: -) covers the required notions, definitions and backgrounds to understand the problem of fault diagnosis of discrete event systems (DES), -) includes multiple illustration examples in various application domains and multiple illustration examples and -) discusses the links between different methods and techniques for the fault diagnosis of DES. The author is very grateful to Brett Kurzman for establishing the contract with Springer Verlag and supporting the author in any organizational aspects. The author would like to thank Rebecca Hytowitz for the text setting including the figures and tables in camera-ready form.

Keywords: Discrete event systems modeling · Discrete event systems diagnosis methods · Diagnosability and co-diagnosability analysis

Contents

Chapter 1
Introduction to the Diagnosis of Discrete Event Systems

1.1 Basic Definitions

A fault can be defined as a non-permitted deviation of at least one characteristic property (feature) of a system, or one of its components, from its normal or intended behavior. This deviation reduces the system performances or its capability to achieve a required function. There are several sources of faults as design faults, wear of system components, wrong operating conditions faults as overloads, maintenance faults, human operators' faults, software and hardware faults etc. A fault can happen regardless the system is in operation or not. Faults can initiate a failure. The latter causes a complete operational breakdown leading to either a permanent or an intermittent interruption of the capability of a system to achieve its required function (see Fig. 1.1). Therefore, a failure can be permanent or intermittent. In the first case, after the occurrence of the failure, the system remains in the faulty conditions indefinitely. In the second case, the failure can appear only during certain periods of time and then it disappears. The number of failures can be single or multiple. They can also be predictable or unpredictable. The latter are characterized as random failures that their occurrence is independent from the operation conditions or the occurrence of other failures. In this book, we will use the two terms, fault and failure, synonymously.

The fault occurrence leads to a change in operating conditions (operation mode) of the system from normal to faulty. This change or transition can occur either abruptly (step-wise faults) or gradually (drift-wise faults) (see Fig. 1.1). In the case of gradual change, the ability (performances) of a system to achieve a required function starts to decrease (degraded behavior) until the failure takes over completely.

Generally, fault diagnosis aims at achieving three tasks: fault detection, fault isolation and fault identification. Fault detection is a binary function which decides whether a fault f has occurred in the system or not. This detection is based on the observation of a difference between the measured and the expected outputs of the system. If a fault has occurred, fault isolation aims at establishing the possible candidate(s) that can explain this difference. The fault candidate represents the system component (plant as tank, actuators as pomp or valve, sensors, etc.) responsible of the occurrence of this fault. Fault identification computes the fault magnitude, its time of occurrence, its importance etc.

M. Sayed-Mouchaweh, *Discrete Event Systems,* SpringerBriefs in Electrical and Computer Engineering, DOI 10.1007/978-1-4614-0031-8_1,

Fig. 1.1 Abrupt and gradual change from normal operating mode towards a faulty one

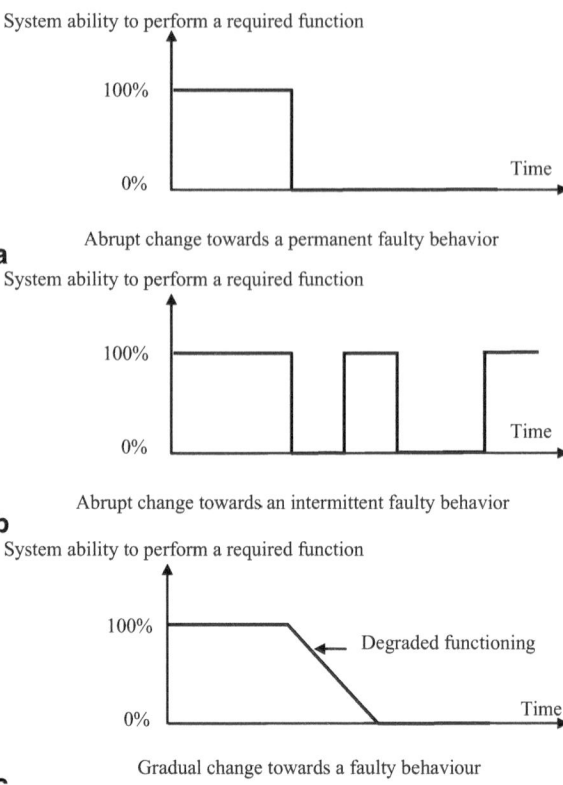

Many diagnosis approaches have been proposed in the literature [1, 4, 6–8, 10–14, 16–18, 20, 22, 24–26, 29, 31, 35, 38, 40, 41, 43, 46, 49, 50–52]. They can be classified according to the fault model compilation in either off-line or on-line. Off-line diagnosis methods assume that the system is not operating in normal conditions but it is in a test bed, i.e., ready to be tested for possible prior failures. The test is based on inputs, e.g. commands, and outputs, e.g. sensors readings, in order to observe a difference between the resulting signals with the ones obtained in normal conditions. In on-line diagnosis, the system is assumed to be operational and the diagnostic module is designed in order to continuously monitor the system behavior, isolate and identify failures. Within these methods, we can distinguish between active diagnosis [40] that uses both inputs and outputs, and passive diagnosis that uses only system outputs. In both cases, they achieve the fault diagnosis by reasoning over differences between desired or expected behavior, defined by a model, and observed behavior provided by sensors. Modeling dynamical systems is a hard and challenging task and the quality of the model impacts significantly the quality of the diagnosis decision.

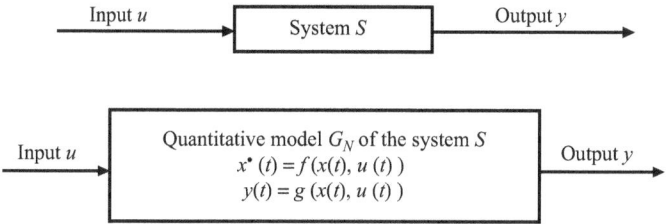

Fig. 1.2 Quantitative model G_N of the normal behavior of a system S described by differential equations

1.2   General Classification of Fault Diagnosis Methods

Generally, two categories of methods are used to build a model of the system behavior: model reasoning and model based approaches. In model reasoning approaches [13, 15, 21, 42], the model about the system behavior is built using an initial human experience, e.g., expert systems, a set of historical data, e.g., pattern recognition and signal processing methods, etc. These approaches are used when the knowledge about the system behavior is incomplete, and thus insufficient to construct a model able to describe all the potential system behavior. Model based approaches [5, 20] establish a mathematical or analytical model about the behavior of a system. In these approaches, depth knowledge about the system behavior is required to build its model. This book focuses on the model based approaches.

The model may be quantitative when the system output can be measured quantitatively precisely. In this case, the model is represented by differential/difference equations, transfer functions, etc. In Fig. 1.2, the system behavior is modeled, described, by the differential equations:

$$x^{\bullet}(t) = f(x(t), u(t)); x(t = 0) = x_0$$
$$y(t) = g(x(t), u(t)) \tag{1.1}$$

Where $x \in \mathbb{R}^n$ is the state vector with the initial state x_0, $y \in \mathbb{R}^r$ is the output vector and $u \in \mathbb{R}^m$ is the input vector. If we take the example of a linear system, the state equations (1.1) are written as follows:

$$x^{\bullet}(t) = Ax(t) + Bu(t)$$
$$y(t) = Cx(t) \tag{1.2}$$

Where A, B and C represents respectively the system, the input and the measurements (output) matrixes.

The diagnosis based on a quantitative model is achieved using parameter identification or residual generation techniques. In parameter identification approaches, a fault causes changes to certain physical parameters and measurements, which in turn will lead to a change in certain model parameters or states. Residual generation approaches exploit the redundancy between the model and the sensor measurements.

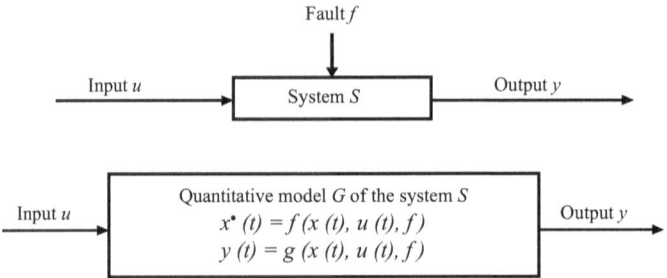

Fig. 1.3 Quantitative model G of the normal G_N and faulty G_F behaviors of system S described by differential equations

They utilize the model differential equations and the available sensors in order to generate relations over observed variables. These relations implicate a subset of faults to be diagnosed. Residuals are then generated based on these relations. They are sensitive (different of zero) to the occurrence of a subset of faults. The challenge of these approaches is to achieve a robust diagnosis to sensor noises, disturbances and model uncertainty, in particular, for multiplicative faults of non-linear systems.

A normal model can be used to confirm a normal behavior faultless case or to detect a fault. However, the fault isolation can not be achieved using only normal behavior. All fault scenarios corresponding to all the possible faults must be considered in the system model. Thus, the differential equations of (1.1) become as follows (Fig. 1.3):

$$x^\bullet(t) = f(x(t), u(t), f); \forall f \in \Sigma_F; x(t=0) = x_0$$
$$y(t) = g(x(t), u(t), f) \tag{1.3}$$

Where Σ_F is the set of all faults that can occur in system S.

The faults are integrated in the model according to their influence on the system behavior. They can be additive or multiplicative. Additive faults can be treated as external unknown inputs. They can affect either the system outputs or its states as sensor or actuator offsets and some kinds of component faults (e.g., leaks in pipes). Multiplicative faults are reflected in system parameter variations ("parametric faults"). They are multiplicative because they multiply themselves with the system state $x(t)$, with its inputs $u(t)$ and its outputs $y(t)$. Therefore they are not as easy to handle as additive faults. Multiplicative faults can, in principle, be approached by additive faults but then they have time-variant coefficients, and they have an effect on the dynamics of the system. The decrease of performance of machines due to their wearing is an example of this type of faults because the initial parameters defined for the operating modes of the system do not represent anymore its actual behavior. If we take a linear system, state equations (1.2) with additive and multiplicative faults can be written, respectively, as follows:

$$x^\bullet(t) = Ax(t) + Bu(t) + Lf_u(t)$$
$$y(t) = Cx(t) + Mf_y(t) \tag{1.4}$$

Fig. 1.4 Tank equipped with
discrete level sensors

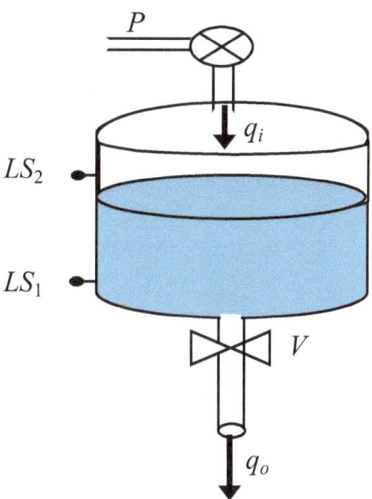

$$x^\bullet(t) = (A + \Delta A)x(t) + (B + \Delta B)u(t)$$
$$y(t) \; = (C + \Delta C)x(t) \tag{1.5}$$

Where f_u denotes the additive fault affecting the system state and f_y is the additive fault affecting the system output. ΔA, ΔB and ΔC represent respectively the multiplicative faults affecting the system, the actuators and the sensors.

The principal advantage of approaches using both normal and fault behaviors, is the precision of the fault isolation. This is crucial for on-line diagnosis in order to help users to make decision about the actions that should be taken at the right time. However, integrating the system behavior in response to a predefined set of faults increases significantly the model size and complexity.

The model is qualitative when the system output is measured qualitatively. When the system changes abruptly its behavior due to the occurrence of a sequence of physical discrete events, then the system is considered as discrete event system [5] and its model is based on a set of events and states. This book focuses on diagnosis of systems modeled as Discrete Event Systems (DES).

The use of DES model is justified by the restriction on the measurability of the system variables describing its behavior. Level or position sensors indicating the level in a tank by three discrete values "Low", "Medium" and "High" or the presence of a piece or a product at a certain position by "Present" and "Absent" are an example of discrete sensors. An event is generated in order to indicate the change of the liquid level in a tank for example from "Low" to "Medium" or from "Medium" to "High", or to indicate the actual sensor reading, e.g., "Sensor reading is High". Thus, DES possess discrete state space and the state transition is "event driven".

Example 1.1 The example of Fig. 1.4 is a tank equipped with pump P supplying water with input flow q_i, valve V with output flow q_o, and two discrete level sensors LS_1 and LS_2. These sensors are positioned at the heights L_1 and L_2 to measure the water level h in the tank. The water level h is the only output (measured variable) of system S.

The normal behavior (change of water level) of system S (the tank) is described by the following differential equations:

$$
\begin{aligned}
x_1^\bullet &= h^\bullet = \frac{q_i - q_o}{s}, \\
x_2^\bullet &= q_o = A\sqrt{2gh}.v(t), \\
x_3^\bullet &= q_i = q_p.p(t); x_0 = (x_{01} = h_0 = 0, x_{02} = q_{o0} = 0, x_{03} = q_{i0} = q_p), \\
y &= x_1 = h
\end{aligned}
\tag{1.6}
$$

Where s is the cross-sectional area of the cylindrical tank, g is the gravitational constant, A is the transversal section of the pipe, and q_p is the flow in the pump. All the state variables are defined in $\mathbb{R}$. The discrete level sensors provide the measures of tank level as follows:

$$
LS_1 = \begin{bmatrix} 1 & \text{if } h(t) \geq L_1 \\ 0 & \text{Otherwise} \end{bmatrix}, \quad LS_2 = \begin{bmatrix} 1 & \text{if } h(t) \geq L_2 \\ 0 & \text{Otherwise} \end{bmatrix}
\tag{1.7}
$$

The inputs of the system are:

$$
u_1 = v(t) = \begin{bmatrix} 1 & \text{if } h(t) > L_2 \\ 0 & \text{Otherwise} \end{bmatrix}, \quad u_2 = p(t) = \begin{bmatrix} 1 & \text{if } h(t) < L_1 \\ 0 & \text{Otherwise} \end{bmatrix}
\tag{1.8}
$$

The DES model of tank S is described by a discrete state space formed by three discrete state variables: water level h in the tank, input Q_i and output Q_o flows. These discrete state variables are defined in the following qualitative domains:

$$
\begin{aligned}
(H) &\in \{\text{"Low","Medium","High"}\}, \\
(Q_i) &\in \{\text{"On","Off"}\}, \\
(Q_o) &\in \{\text{"Closed","Open"}\}
\end{aligned}
\tag{1.9}
$$

Figure 1.5 shows the continuous water level signal and the corresponding discrete state variables, events and level sensors.

The discrete tank level can be described by two Boolean state variables: LS_1 and LS_2. Similarly, the state of pump P (valve V) can be described by a Boolean variable indicating if the pump (the valve) works with its maximum power, $P = 1$ ($V = 1$), or it does not work, $P = 0$ ($V = 0$). Each transition of the state of the tank level, the pump or the valve generates an event. The set of generated events are shown in Table 1.1. Each event can correspond to either a rising edge $\uparrow x$, transition of the discrete state value x from 0 to 1, or failing edge $\downarrow x$, transition of the discrete state value x from 1 to 0.

In order to integrate the faulty behavior, the set of all potential faults must be defined as we can see in Table 1.2 for the example of the simple tank of Fig. 1.3. Thus, five models will be built: the one corresponding to the normal behavior G_N and the faulty behaviors $G_{F_i}, i \in \{1, ..., 4\}$ corresponding to the behavior in response

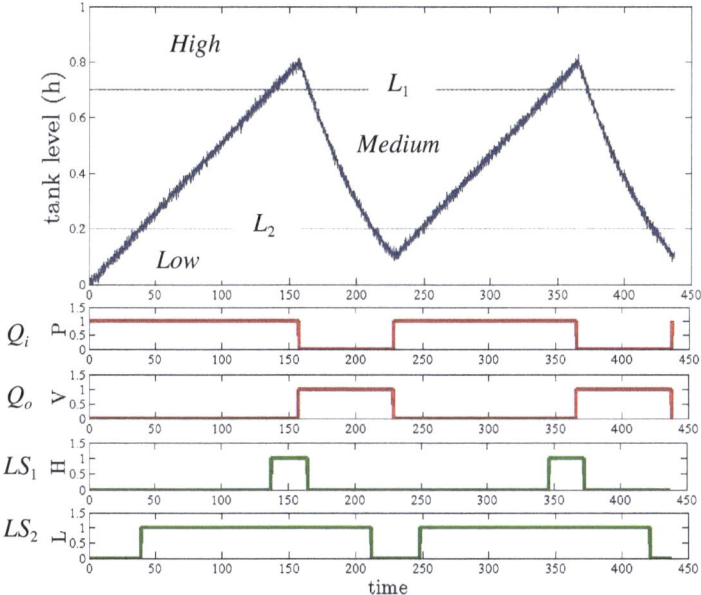

Fig. 1.5 Discrete state variables and the generated events for the example of simple tank of Fig. 1.4

to a fault f belonging to each fault partition $\Pi_{F_i}, i \in \{1, ..., d = 4\}$. Π_{F_i} is the partition, i.e., subset, of faults associated with the fault label F_i. Each fault partition corresponds to some kinds of faults in a system component which have the same effect according to either the configuration or maintaining procedure. Σ_Π is the set of all fault partitions which can appear in the system: $\forall i \{1, ..., d\}, \Pi_{F_i} \in \Sigma_\Pi = \left\{ \Pi_{F_1}, ..., \Pi_{F_d} \right\}$.

1.3   Fault Diagnosis of Discrete Event Systems

Many approaches have been proposed to achieve the diagnosis. They are generally based on the use of a model of nominal (desired) behavior G_N of the system and a model for each faulty behavior G_{F_i} in response to a specified fault $f \in \Pi_{F_i}, i \in \{1, ..., d\}$, where Π_{F_i} is the partition, i.e., subset, of faults associated with the fault label F_i (see Fig. 1.6).

The DES fault diagnosis methods can be classified according to several criteria:

- *Model description tool.* In general, two description tools can be used to build a model: automata and Petri nets [5]. Thus, two categories of diagnosis methods can be distinguished: automata [10, 12, 16–18, 22, 25, 26, 29, 31, 35, 36, 38, 39, 41, 46, 51, 52] and Petri nets [6, 14, 18, 24, 37] based methods. In these two

Table 1.1 Generated events for the example of the simple tank of Fig. 1.4

Events	signification
$\uparrow LS_1$	Raising edge event indicating that the level of liquid in the tank is "Medium"
$\downarrow LS_1$	Failing edge event indicating that the level of liquid in the tank is "Low"
$\uparrow LS_2$	Raising edge event indicating that the level of liquid in the tank is "High"
$\downarrow LS_2$	Failing edge event indicating that the level of liquid in the tank is "Medium"
$\uparrow Q_i$	Raising edge event indicating that the pump is turned "On"
$\downarrow Q_i$	Failing edge event indicating that the pump is turned "Off"
$\uparrow Q_o$	Raising edge event indicating that the valve is "Open"
$\downarrow Q_o$	Failing edge event indicating that the valve is "Closed"

categories, the system model and the diagnosis module are constructed using, respectively, automata or Petri nets.

- *Fault representation and inference.* Two main categories of approaches are used to represent and to infer a fault occurrence: event-based-diagnosis [17, 22, 23, 26, 29, 38] and state-based-diagnosis approaches [26, 50, 51]. In the first category, the occurrence of a fault is detected and isolated based on the observation of event sequences or/and their date of occurrence. The fault diagnosis in the second category of approaches is performed by processing the state output transition sequences.

- *Diagnosis processing structure.* Three main processing structures are used to calculate the fault diagnosis decision: centralized [22, 23, 27, 39], decentralized [4, 12, 31, 35, 36, 41, 49, 52] and distributed [10, 16, 18, 24, 46] structures. In the centralized structure (Fig. 1.7a), the diagnosis is calculated using one global diagnosis module, called diagnoser. The latter is constructed using one global model of the system. In decentralized structure (Fig. 1.7b), the diagnosis is performed based on a set of local diagnosers. Each local diagnoser computes a local diagnosis decision based on its partial observation about the whole system. A

Table 1.2 Definition of faults and their labels for the example of a simple tank of Fig. 1.4

Fault partition	Faults	Fault definition	System component	Fault label
Π_{F_1}	f_1	Valve V stuck-on	Valve V	F_1
	f_2	Valve V stuck-off		
Π_{F_2}	f_3	Pump P failed-on	Pump P	F_2
	f_4	Pump P failed-off		
Π_{F_3}	f_6	Sensor LS_1 stuck-on	Sensor LS_1	F_3
	f_7	Sensor LS_1 stuck-off		
Π_{F_4}	f_8	Sensor LS_2 stuck-on	Sensor LS_2	F_4
	f_9	Sensor LS_2 stuck-off		

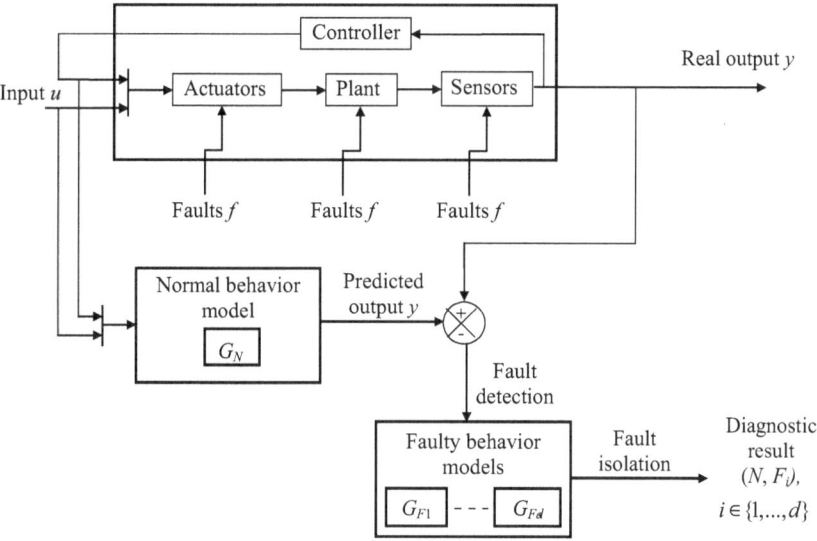

Fig. 1.6 General scheme for fault diagnosis of discrete event systems

coordinator receives the local diagnosis decisions in order to provide the final diagnosis decision. No direct communication among the local diagnosers is allowed. Only a limited communication among them through the coordinator is permitted. In the third structure (Fig. 1.7c), the system is divided into subsystems. Each subsystem observes only its own part of the global model. A local diagnoser is associated to each subsystem in order to perform diagnosis locally. This diagnosis computation is based on the local model and the information communicated directly to it by the other local diagnosers through a communication protocol. The information exchanged among local diagnosers is used by them to update their own information. This information update compensates the local diagnosers partial observation.

This book focuses on the centralized and decentralized diagnosis methods based on the use of automata.

1.4   Diagnosability and Co-Diagnosability Notions

The diagnosability property can be defined as the capacity of a diagnoser to infer the occurrence of a fault f as well as its fault partition $f \in \Pi_{F_i}, i \in \{1, ..., d\}$ based on the observation of the sequences of observable events. The diagnosability property can be verified using a diagnosability notion. Several diagnosability notions exist in the literature [8, 22, 26, 28, 32, 33, 39, 47, 50, 52]. Generally, they differ according to:

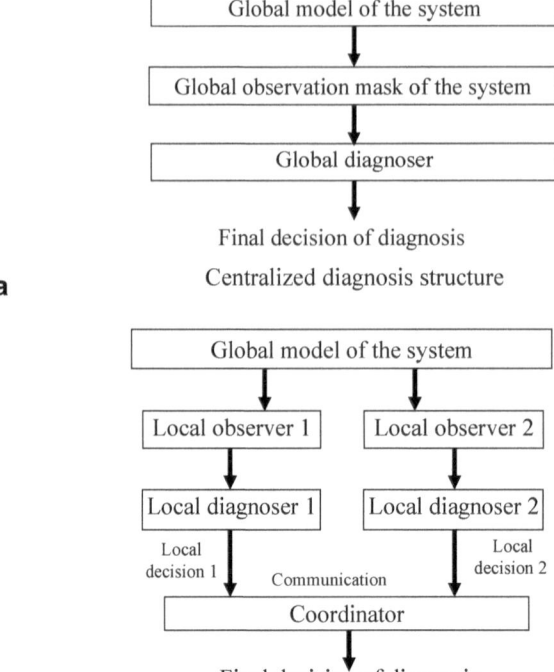

a   Centralized diagnosis structure

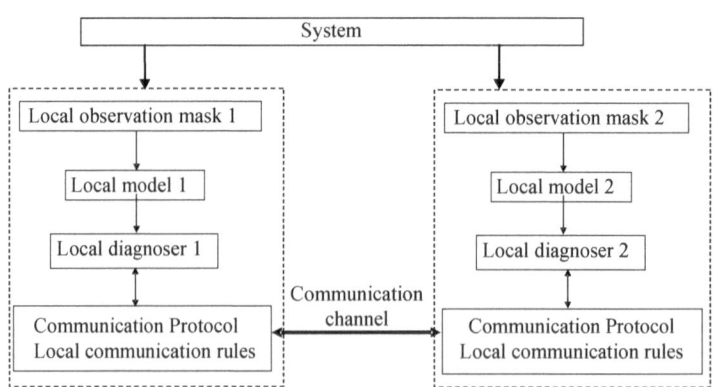

b   Decentralized diagnosis structure in the case of a system composed of two sites

c   Distributed diagnosis structure in the case of a system composed of two sites

Fig. 1.7 Diagnosis decision structures

- the diagnosis decision structure: diagnosability notion [22, 26, 39] for centralized structure and co-diagnosability notion [12, 32, 45, 46] for decentralized and distributed structures;
- the fault occurrence inference: event-based [22, 26, 39] and state-based [26, 41, 50] diagnosability notions;
- the connections between the system components: local [44, 52], independent [41, 44], joint [46] and decentralize [12, 32] diagnosability notions.

This book focuses on the event based diagnosability and co-diagnosability of centralized and decentralized DES.

Chapter 2
Centralized Diagnosis of Discrete Event Systems

2.1 Introduction

As we have seen in Chap. 1, centralized diagnosis requires one centralized model of the targeted system associated with one centralized diagnoser (see Fig. 1.7a). The latter collects observations about the system behaviors in one central point. Then, it treats these observations in order to make decision about the occurrence of a fault and its responsible elements (e.g., actuator, sensor, etc.). Examples of a centralized diagnosis structure can be found in [22, 23, 27, 39] and the references therein.

This chapter will focus on two significant event-based approaches of centralized diagnosis of DES: diagnoser and supervision pattern approaches. In these approaches, the system is represented (modeled) by an automaton in order to achieve or to solve the problem of diagnosis. The events, which change the system state, are divided into two disjoint sets: observable and unobservable events. The faults are considered as unobservable events. Thus, the automaton is nondeterministic with unobservable transitions. All the information relevant to the diagnosis problem of a system is captured in the framework of events generated by this system. Therefore, diagnosis problem is solved by observing the set of observable event sequences or strings (words). In other words, the occurrence, if any, of failure events, is inferred using the set of generated words containing only observable events.

Model G, generating formal language L, can represent either the normal and faulty behaviors of the system (diagnoser approach) or only specified faulty behaviors (supervision patterns). In the first case, the nondeterministic automaton G, representing the system under consideration, is converted into deterministic one by considering only the observable events (observable transitions). The resulting deterministic automaton is called observer $Obs(G)$. Each node of $Obs(G)$ contains the states that the system can be in in response to the occurrence of an observable event sequence. Then, the diagnosis problem is solved by building a diagnoser $Diag(G)$. The latter is an observer but information about the occurrence or not of each fault is added to each of its nodes. This information is represented by a label indicating whether the system is fault-free (N) or fault f, belonging to fault partition Σ_{F_i} occurred. In the second case, a faulty behavior in response to the occurrence of fault f is modeled as a set of partial observable trajectories (traces or event sequences) that one wants

M. Sayed-Mouchaweh, *Discrete Event Systems,* SpringerBriefs in Electrical and Computer Engineering, DOI 10.1007/978-1-4614-0031-8_2, © Author 2014

Fig. 2.1 System model G

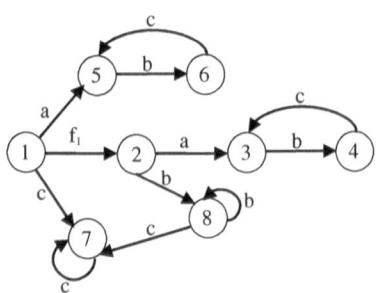

to recognize their occurrence. The diagnosis of the occurrence of f is achieved by matching between the real behavior of the system and the compiled faulty behavior.

In both cases, the following assumptions hold:

(A1) Language L generated by model G is live. This means that each state of G has a transition.

(A2) There does not exist in G any cycle of unobservable events. A2 ensures that G does not contain sequences of unobservable events whose length can be infinite.

(A3) Faults cause a distinct change in the system status but do not necessarily bring the system to a halt.

2.2 Diagnosability Notion

The diagnosability notion is based on the fact that a system is diagnosable if and only if any pair of faulty/non-faulty behaviors can be distinguished by their projections to observable behaviors. Before defining the necessary and sufficient conditions for a system to be diagnosable, let us consider the following definitions.

Definition 2.1 Language L generated by system model G is the set of all the event sequences or trajectories u that the system can execute.

Example 2.1 Let us take the example of model G represented in Fig. 2.1. The language generated by G is: $L = \{ f_1a(bc)^*, a(bc)^*, f_1bb^*, cc^*, f_1cc^* \}$.

Definition 2.2 Let Σ be the set of all events that can be generated by a system S. Let Σ^* denote the set of all event sequences that can be formed using the events in Σ. Let Σ_o denotes the set of observable events generated by system S. Then, Σ_o^* is the set of all event sequences that can be formed using observable events in Σ_o. The projection function $P : \Sigma^* \rightarrow \Sigma_o^*$ allows erasing all unobservable events in an event sequence.

Example 2.2 Let us take the example of Fig. 2.1 where: $\Sigma = \{a, b, c, f_1\}$ and $\Sigma_o = \{a, b, c\}$. The projection $P(u)$ of event sequence $u = f_1a(bc)^* \in \Sigma^*$ is $a(bc)^* \in \Sigma_o^*$ since f_1 is unobservable event.

Fig. 2.2 Diagnoser for the
example of Fig. 2.1

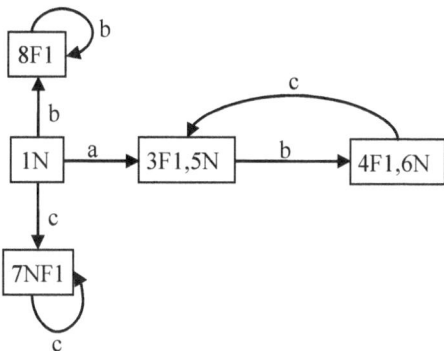

Definition 2.3 A diagnoser state q_D contains a set of pairs (x, l) where x is a state of system model G and l is a fault partition label or normal label N. Diagnoser state q_D is said to be F_i-certain state if for all the pairs (x, l), l is equal to F_i.

Diagnoser state q_D is said to be F_i-uncertain if it contains two pairs (x_1,l_1) and (x_2,l_2) where $l_1 = F_i$ and $l_2 \neq F_i$.

Diagnoser state q_D is ambiguous if there is at least two pairs (x_1,l_1) and (x_2,l_2) where $x_1 = x_2$ and $l_1 = F_i$ and $l_2 \neq F_i$.

Example 2.3 Let us take the example of Fig. 2.1. Figure 2.2 shows the corresponding diagnoser D. Diagnoser state $q_D = (8F1)$ is F_1-certain state since it contains only the fault label F_1. Diagnoser state $q_D = (3F1,5N)$ is F_1-uncertain state since it contains two states with fault label F_1 and the normal label N. Diagnoser state $q_D = (7F1,7N)$ is F_1-ambiguous state since it contains the state model 7 which has the normal and F_1 labels.

Definition 2.4 F_i-indeterminate cycle in diagnoser D is a cycle composed exclusively of F_i-uncertain states. It indicates the presence in L of two event sequences u_1 and u_2 such that they both have the same observable projection, $P(u_1) = P(u_2)$, and u_1 contains a failure event while u_2 does not.

Example 2.4 Cycle $(bc)^*$ in Fig. 2.2 is F_1-indeterminate cycle since it corresponds to two cycles in the model G of Fig 2.1; the first cycle is formed by states with normal label and the second cycle by states with F_1 fault label.

Definition 2.5 Two conditions are required for a system to be F_i diagnosable [39]:

1. after the occurrence of fault f belonging to the fault partition of label F_i, the diagnoser must visit or reach an F_i-certain diagnoser state within a finite number of observable events,
2. the diagnoser must not contain any indeterminate cycle.

In the next, the diagnosability notion will be studied using several examples.

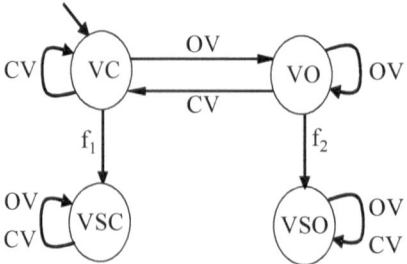

VC: Valve in closed state
VO: Valve in Opened state
VSC: Valve stuck close
VSO: Valve stuck open
OV: Open valve controllable event
CV: Close valve controllable event
f1: Valve stuck close fault event
f2: Valve stuck open fault event

Fig. 2.3 Normal and faulty behaviors for the valve model

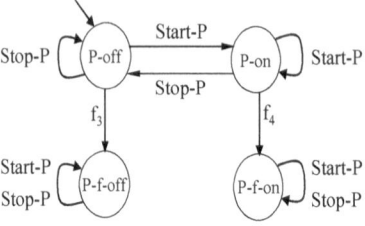

P-off: Pump in on state
P-on: Pump in off state
P-f-off: Pump failed off
P-f-on: Pump failed on
Start-p: Start pump controllable event
Stop-p: Stoppump controllable event
f3: Pump failed off fault event
f4: Pump failed on fault event
F2: Fault label indicating the occurrence
   of f3 or f4

Fig. 2.4 Normal and faulty behaviors for the pump model

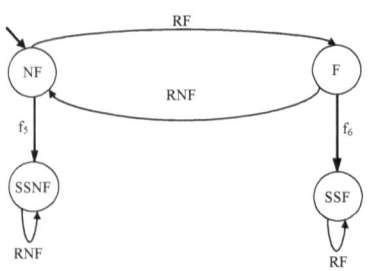

NF: Sensor in non-flow state
F: Sensor in flow state
SSNF: Sensor stuck at no-flow state
SSF: Sensor stuck at flow state
RF: Sensor reading to be flow uncontrollable event
RNF: Sensor reading to be no-flow uncontrollable event
f_5: Sensor stuck in no-flow state fault event
f_6: Sensor stuck in flow state fault event
F_3: Fault label indicating the occurrence of f_5 or f_6

Fig. 2.5 Normal and faulty behaviors for the flow sensor model

2.3 Diagnoser Approach

In Diagnoser approach [39], the diagnosis problem is the task of assigning to each observed string (word) of events a diagnosis state with one of the following status: "normal", "faulty" or "uncertain". The uncertainty can be reduced by continuing to make observations.

The diagnoser approach works as follows. The system to be diagnosed is supposed to be composed of *n* individual components. Typically, these components consist of equipment (actuators and sensors) and controllers. However, they can also represent

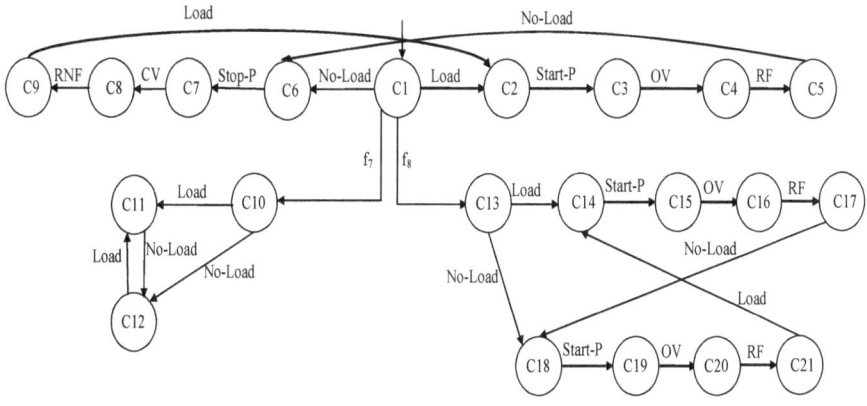

f₇: Controller failed off fault event
f₈: Controller failed on fault event
F₄: Fault label indicating the occurrence of f₇ or f₈

Fig. 2.6 Normal and faulty behaviors for the controller model

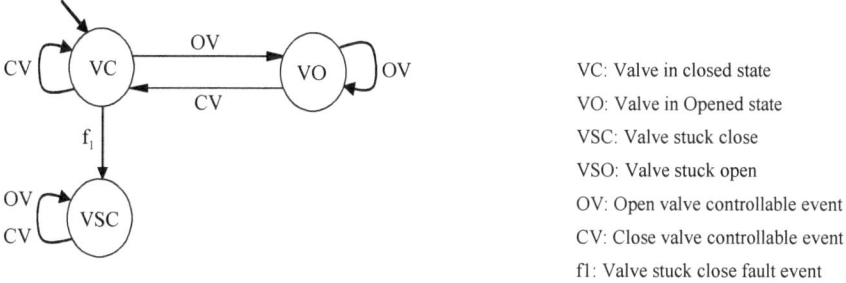

VC: Valve in closed state
VO: Valve in Opened state
VSC: Valve stuck close
VSO: Valve stuck open
OV: Open valve controllable event
CV: Close valve controllable event
f1: Valve stuck close fault event

a   Model of valve V

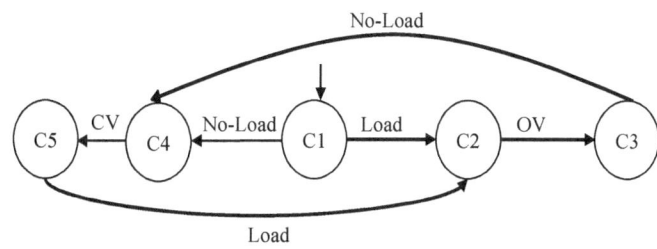

b   Controller model

Fig. 2.7 Components models for Example 2.6

the process itself (e.g., a tank). The construction of the diagnoser is based on the following steps [39]:

1. **Step 1**: Build the Finite State Machine (FSM) models for the system components. The model representing the normal behavior of each component is firstly built.

Fig. 2.8 Composite model of
the example of Fig. 2.7

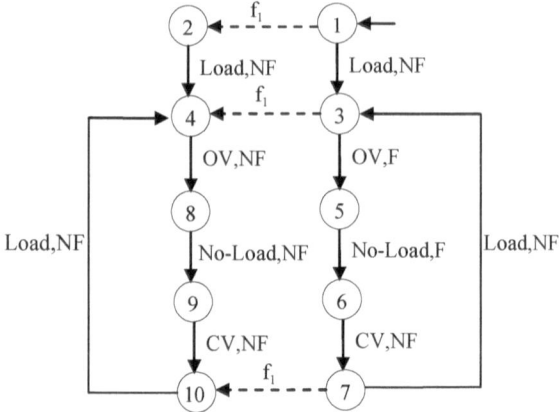

Fig. 2.9 Diagnoser for the
example of Fig. 2.8

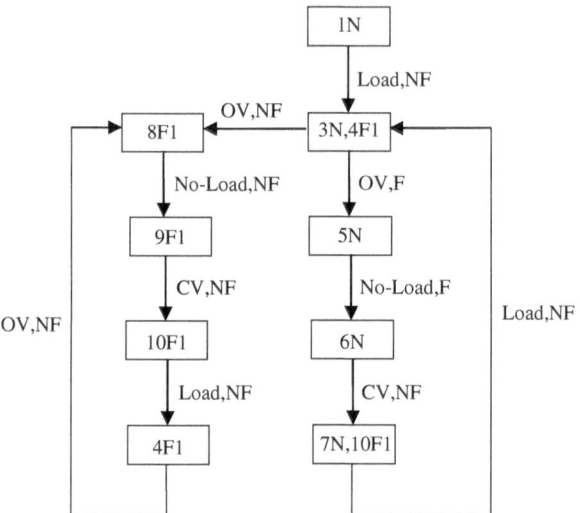

Then, the failed behaviors corresponding to the occurrence of a predefined set of
faults are integrated in the component normal FSM model.

2. **Step 2**: Obtain the global model of the system by applying the standard
   synchronous composition operator between the components individual models.
3. **Step 3**: Generate the global sensor map that lists the discrete sensor readings for
   each state of the global model built in step 2. Convert the sensor readings into
   observable event framework as follows. Reading of the sensor output is considered
   as an observable event after immediately the execution of a control command.
   Each transition of the global model (built in step 2) will be renamed by adding
   the corresponding sensor reading event to the original observable event. In the
   case that the execution of a command leads to a change in the sensor reading, this
   sensor change is considered as an observable event. Thus, a transition associated
   with the sensor change reading is added to the model.

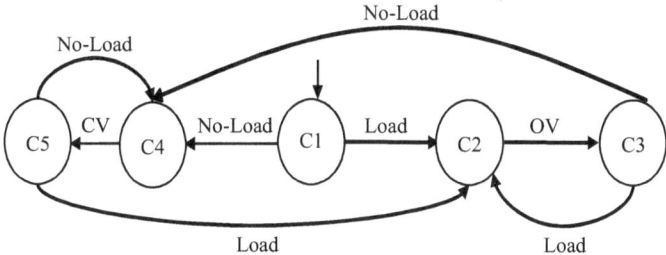

Fig. 2.10 New controller model for the example of Fig. 2.7

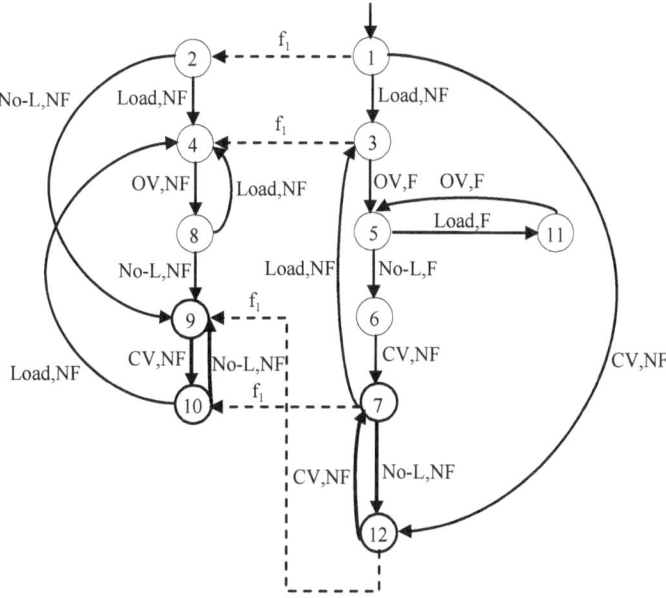

Fig. 2.11 Composite model for Example 2.7

4. **Step 4**: Construct the diagnoser based on the use of the global model. The diagnoser contains only observable events. Each of its states includes one or more of labels indicating a fault-free (N) or the occurrence of faults belonging to predefined fault partitions: $\Pi_{F_i}, \forall i \in \{1, ..., d\}$.

In the next, this approach is explained using a simple example of a valve, a pump, a discrete flow sensor and a controller.

Example 2.5 Let us take the following example extracted from [39]. This example consists of a pump, a valve and a controller. Let us suppose that the system is equipped with one sensor to indicate the presence of a flow at the output of the valve. This sensor has one of the two outputs: F to indicate the presence of a flow and NF to indicate no flow. Let us consider the behavior of the valve V. The valve can be in one of two different states: closed 'VC' and opened 'VO'. Let us assume that the valve can fail due to a stuck-at-on fault or to a stuck-at-off fault. 'VSO' represents the state

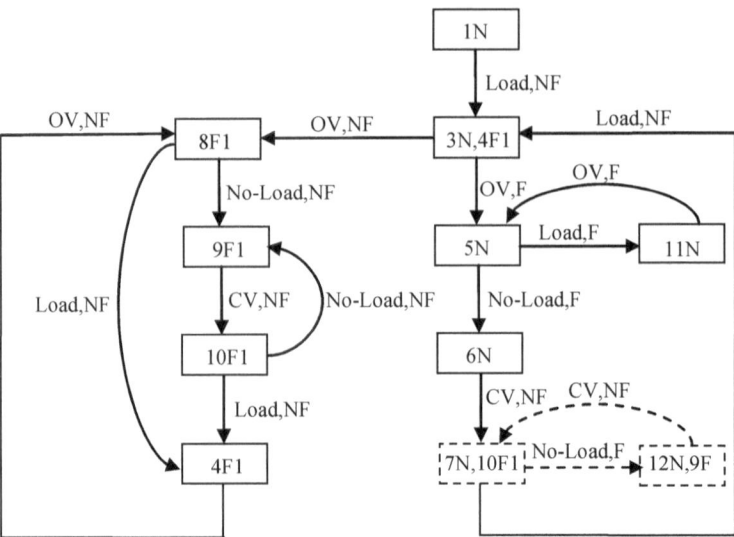

Fig. 2.12 Diagnoser for the example of Fig. 2.11

Fig. 2.13 Composite model
for Example 2.8

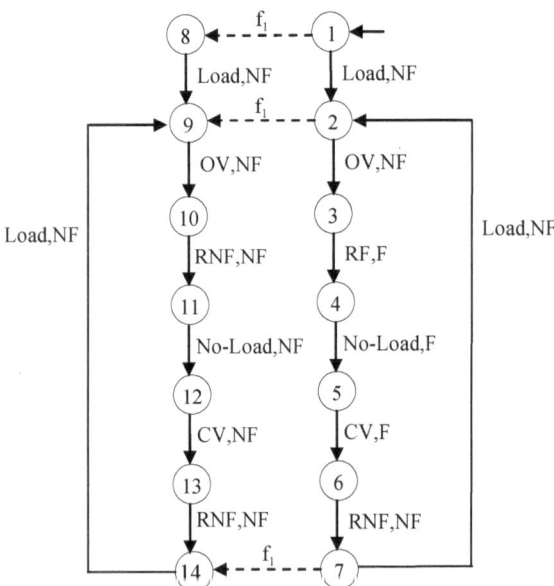

of valve stuck-open and 'VSC' represents the state of valve stuck-closed. In Fig 2.3 the FSM modeling the normal and faulty behaviors of valve V is depicted. The system events are: 'OV' that represents the command of opening the valve, 'CV' that represents the command of closing the valve, f_1 and f_2 that represent respectively the stuck-close and stuck-open events. In a normal behavior, the valve is in 'VC' and can change its state when one of commands 'OV' and 'CV' is sent by the controller. The valve may fail. Either a stuck-close fault (f_1) or a stuck-open fault (f_2) can occur.

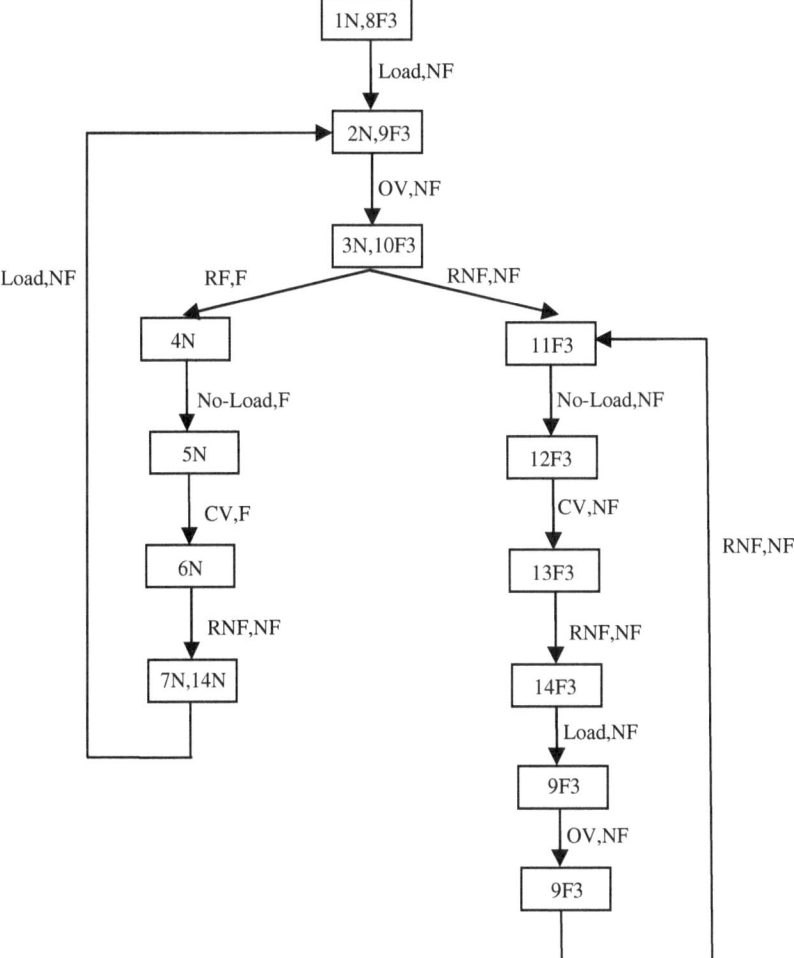

Fig. 2.14 Diagnoser for the example of Fig. 2.13

After the occurrence of one of these faults, both commands (open and/or close the valve) do not change the state of the valve. Events '*OV*' and '*CV*' are controllable and thus observable events, while f_1 and f_2 are faulty and unobservable events and we have to infer their occurrence on the basis of observable event sequences. Figure 2.4 and 2.5 show, respectively, the FSM modeling the normal and faulty behaviors of the pump and the flow sensor. Figure 2.6 shows the command issued by the controller. In normal operation mode, when there is a load in the system, the controller responds by starting the pump and opening the valve. When there is no load anymore in the system, the controller stops the pump and closes the valve. When the controller fails off (i.e., when event f_7 occurs), it does not sense the presence of load on the system and therefore it does not send any of the above commands. While, when the controller

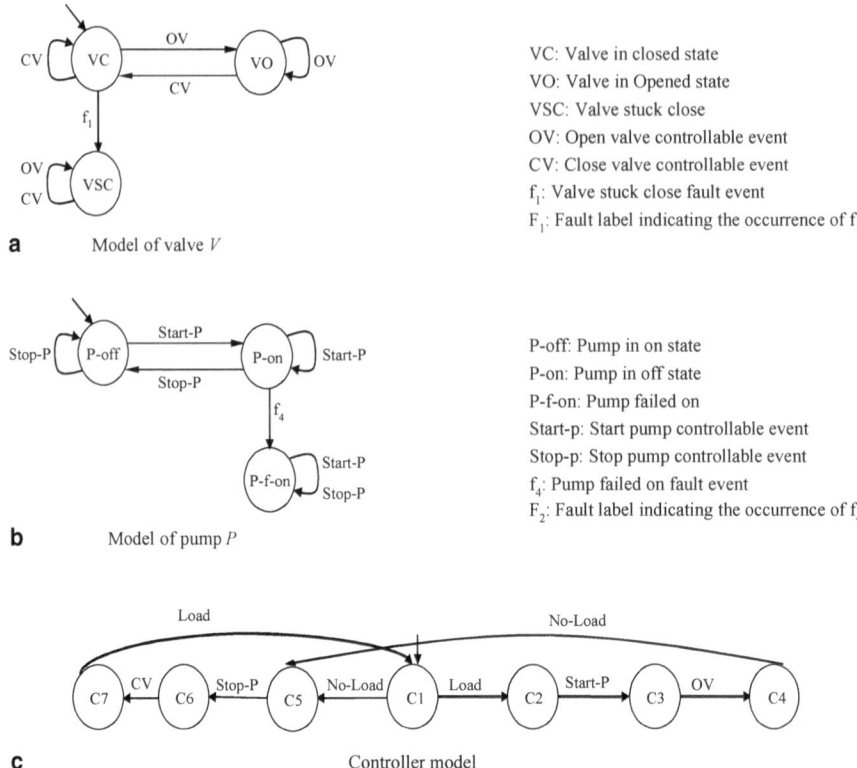

a   Model of valve V

VC: Valve in closed state
VO: Valve in Opened state
VSC: Valve stuck close
OV: Open valve controllable event
CV: Close valve controllable event
f_1: Valve stuck close fault event
F_1: Fault label indicating the occurrence of f_1

b   Model of pump P

P-off: Pump in on state
P-on: Pump in off state
P-f-on: Pump failed on
Start-p: Start pump controllable event
Stop-p: Stop pump controllable event
f_4: Pump failed on fault event
F_2: Fault label indicating the occurrence of f_4

c   Controller model

Fig. 2.15 Components and controller models for Example 2.9

fails on (when f_8 occurs), it assumes a presence of load and consequently it issues the command sequence $< Start\text{-}P > < OV >$ regardless of whether a load is actually present or not. It is supposed that the controller does not fail during operation. If it does fail, the fault occurs at the start of operation.

Example 2.6 In order to facilitate the example of pump-valve (Example 2.5), let us suppose that only the valve can fail in the stuck close failure mode (VSC). Therefore, failure state 'VSO' in the valve model of Fig. 2.3 can be removed. Let us suppose that the pump is always in its on state. The pump model can thus be removed. Since, we are not interested in diagnosing sensor faults; the model of the sensor can be also removed as well as its events from the controller model. Therefore, we have the following models for the valve and the controller shown in Fig. 2.7.

The composite model (obtained by the synchronous composition of the system component models of Fig. 2.7) is depicted in Fig. 2.8. The sensor reading $\{NF, F\}$, considered as an observable event, will be added to this composite model by associating it to each command event. To achieve that, the flow sensor map must be constructed. The sensor map helps to associate to each observable event (typically

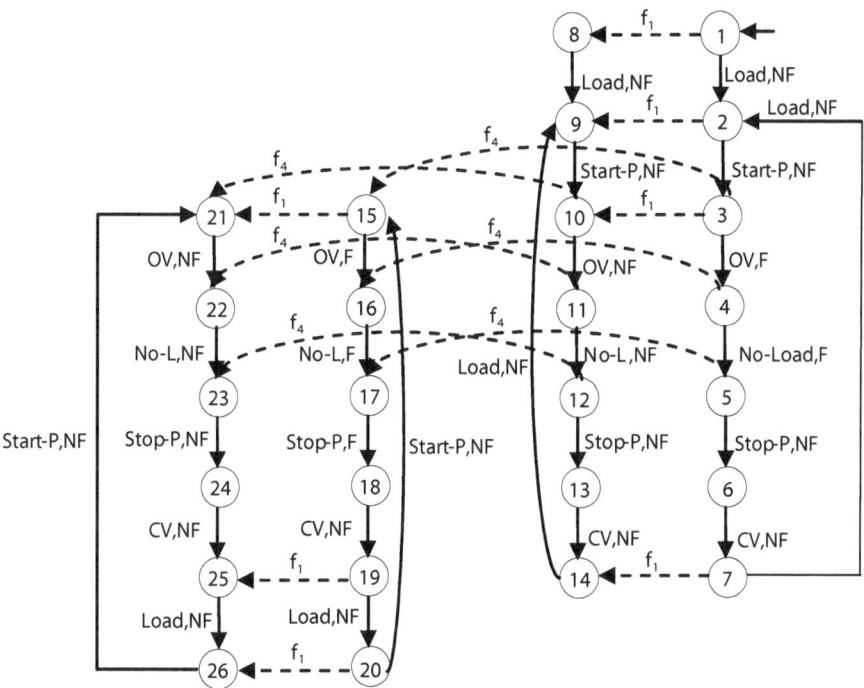

Fig. 2.16 Composite model of the components and controller models of Fig. 2.15 for Example 2.9

commands issued by the controller) the sensor reading. The map indicates the sensor output according to each state of the composite model. This sensor map is independent of the command (the controller) and depends only on the system composite model state. The flow sensor map for the example of Fig. 2.7 is defined as follows: $h(VC,.)=NF$; $h(VO,.)=F$, $h(VSC,.)=NF$. The point (.) indicates that the sensor reading for the corresponding state of the composite model (formed by the combination of the system components models) is independent of the command that will be issued at this state.

As an example, the sensor reading for the state 1 corresponding to Valve Closed state (*VC*), Pump on state (*P-on*) and to controller state $C1$ is No Flow (*NF*). Thus, the command (Load) issued by the controller at this state 1 will be associated to sensor reading *NF*.

The diagnoser (see Fig. 2.9) for the example of Fig. 2.7 is constructed based on the use of its composite model of Fig. 2.8. The diagnoser contains only the observable events. Each diagnoser state contains the composite model states associated with a label indicating the occurrence of a fault (with its fault partition label) or not (fault-free case with label *N*). When a diagnoser state contains two composite model states with two different labels, this means that this diagnoser state can be reached due to the occurrence of some unobservable events. At least one of these unobservable events is a failure event. If one of these labels is the "*N*" label, then the diagnoser

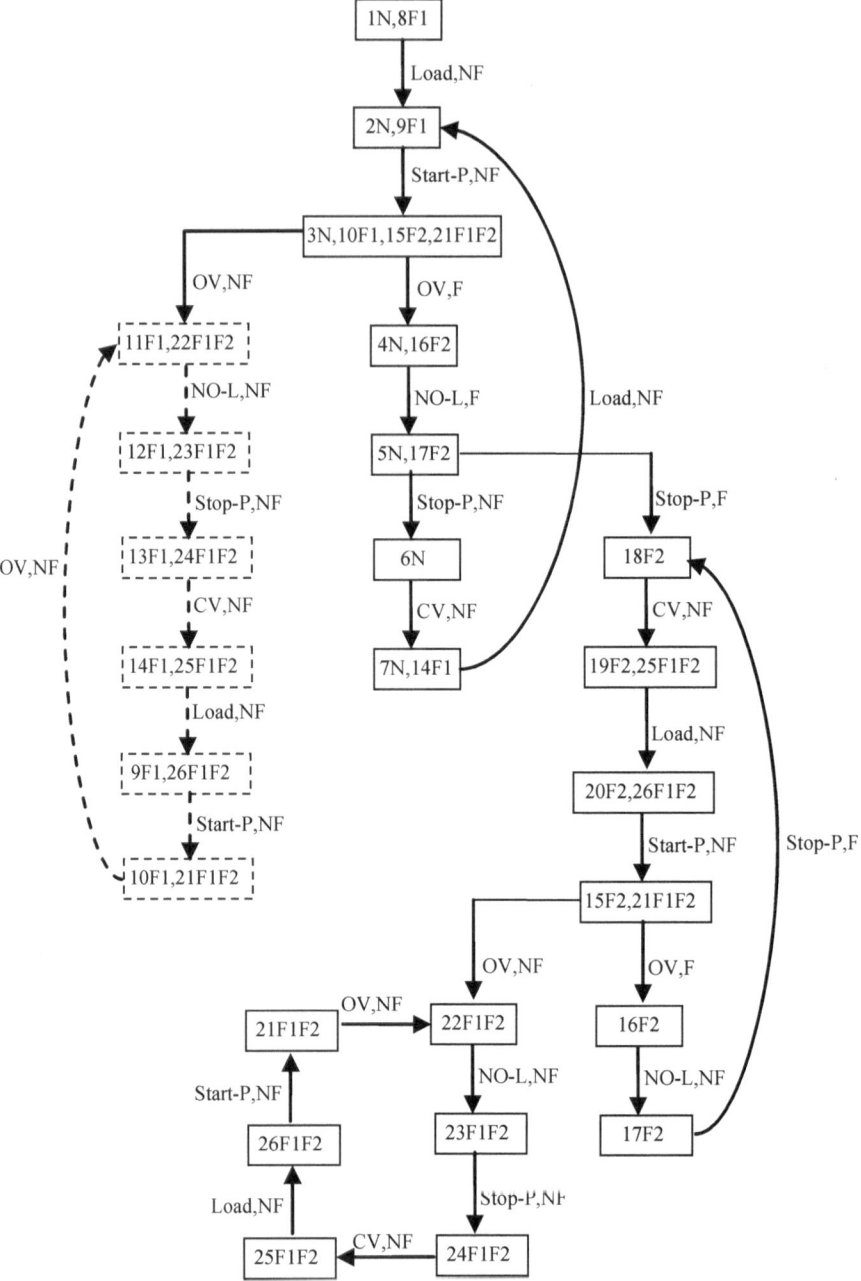

Fig. 2.17 Diagnoser for the example of Fig. 2.16

Fig. 2.18 Global model G for
Example 2.10

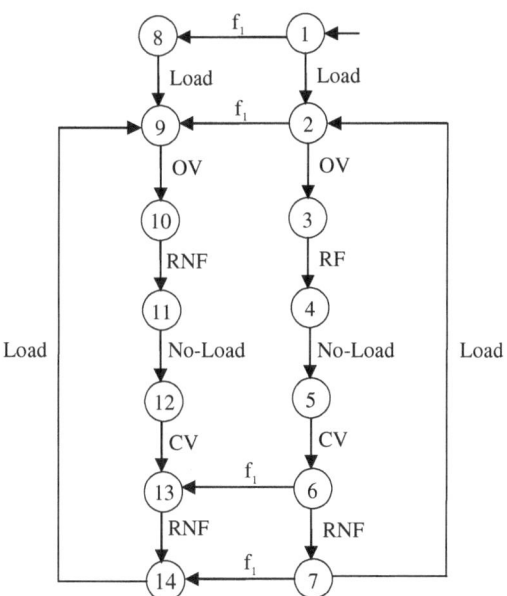

state is uncertain. An uncertain diagnoser state (see Definition 2.3) means that the diagnoser is unable to decide with certainty whether the system works in normal conditions or a fault has occurred. A diagnoser certain state contains only one fault label in the case of single fault occurrence scenarios. In this state, the diagnoser can decide with certainty the occurrence of a fault belonging to a specified fault partition. Figure 2.9 shows the diagnoser for the example of Fig. 2.8.

The diagnoser of the example of Fig. 2.9 is F_1 diagnosable because (see Definition 2.5):

1. the diagnoser reaches F_1-certain state (8F1) within a finite number of observable events (see Definition 2.3),
2. it does not contain any F_1-indeterminate cycle (see Definition 2.4).

Example 2.7 Let us take the same example of Fig. 2.7 but with the controller model depicted in Fig. 2.10 and let us construct the corresponding diagnoser. Based on the same reasoning used for Example 2.6, the composite model and its corresponding diagnoser can be constructed as we can see respectively in Fig. 2.11 and 2.12.

The diagnoser of Fig. 2.12 is not F_1 diagnosable (see Definition 2.5) because the diagnoser contains F_1-indeterminate cycle $(< No\text{-}Load, F > < CV, NF >)^*$, the dotted cycle in Fig. 2.12 (see Definition 2.4). Indeed, in the system model, Fig. 2.11, there are two cycles, in bold in Fig. 2.11. The first cycle (states 7 and 12) contains fault free fault states while the second cycle (states 10 and 9) comprises faulty states. These two cycles have the same observable projection of the event sequence $(< No\text{-}Load, F > < CV, NF >)^*$.

Example 2.8 Let us consider for Example 2.5, that only the sensor can fail in sensor stuck at NF state (reached due to the occurrence of fault f_5). The state corresponding

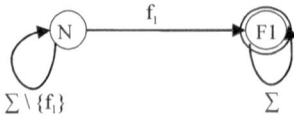

Fig. 2.19 Supervision pattern Ω as a Labeled Transition System (LTS) for the example of Fig. 2.18. Ω represents any faulty behavior leading to the stuck-close failure state of the valve

to the sensor stuck at F state can be removed from the sensor model of Fig. 2.5. The failure states of the valve of Fig. 2.3 can be also removed since its faults are not considered. Let us also consider that the pump is always in its on-state. Thus, there is no need to the use the Pump model. Let us consider the controller model of Fig. 2.10. Figure 2.13 shows the composite model combining the valve, the sensor and the controller models. Based on this composite model, the corresponding diagnoser can be constructed as we can see in Fig. 2.14.

The diagnoser of Fig. 2.14 is F_3 diagnosable because (see Definition 2.5):

1. the diagnoser reaches F_3-certain state (11F3) within a finite number of observable events (see Definition 2.3),
2. it does not contain any F_3-indeterminate cycle (see Definition 2.4).

Example 2.9 The diagnoser approach does not make any assumptions on the number of failures; it is general enough to accommodate multiple system failures. This example handles the case of multiple failures from different fault types.

Fig. 2.20 $G_\Omega = G x \Omega$ between model G (Fig. 2.18) and supervision pattern Ω Fig. 2.19 for Example 2.10

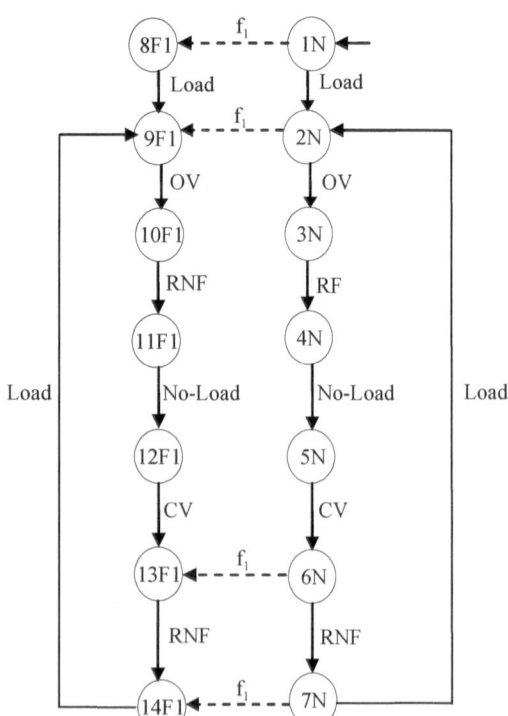

Fig. 2.21 $OBS(G_\Omega)$ for the
example of Fig. 2.20

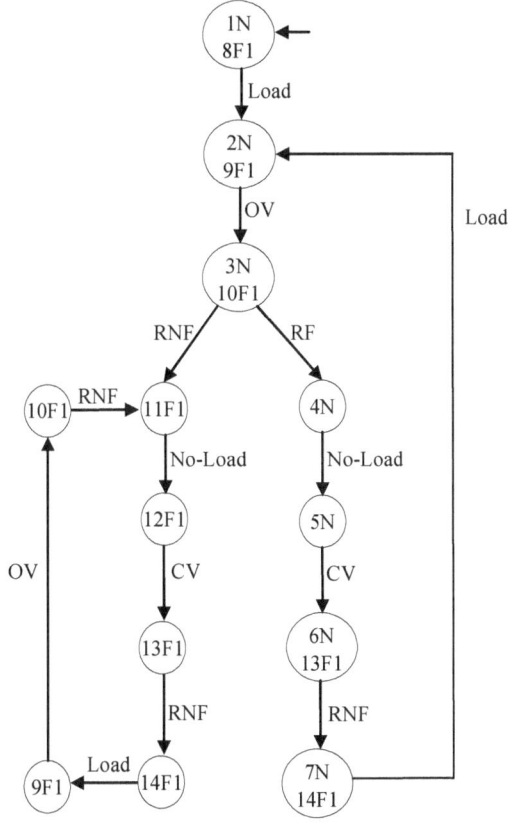

Let us take the pump-valve example. Let us consider that the valve can fail at the stuck-close failure mode (occurrence of fault f_1 of type F_1), as it is depicted in Fig. 2.15a, and the pump can fail at failed-on failure mode (occurrence of fault f_4 of type F_2) as we can see in Fig. 2.15b. The controller model for this example is shown in Fig. 2.15c. We consider in this example that both the valve and the pump can fail; the multiple failures scenario can thus happen. Figure 2.16 shows the global model G, resulting from the synchronous composition of the models of Fig. 2.15. Figure 2.17 shows the corresponding diagnoser. We can notice the case of the occurrence of multiple failure modes (occurrence of both f_1 and f_4) represented by fault labels F_1F_2.

The diagnoser of Fig. 2.17 is not F_1F_2 diagnosable because the diagnoser contains F_1F_2 indeterminate cycle, ($< No\text{-}L, NF > < Stop\text{-}P, NF> < CV, NF > < Load, NF > < Start\text{-}P, NF > < OV, NF >)^*$, shown as dotted cycle in Fig. 2.17. Indeed, the occurrence of pump failed-on failure (F_2) hides the occurrence of valve stuck-closed fault (F_1); thus the diagnoser cannot infer with certainty their joint occurrence. As we have seen in Example 2.6, the valve stuck-closed fault when it occurs alone can be diagnosed with certainty (see Fig. 2.8).

Fig. 2.22 Global model G for
Example 2.11

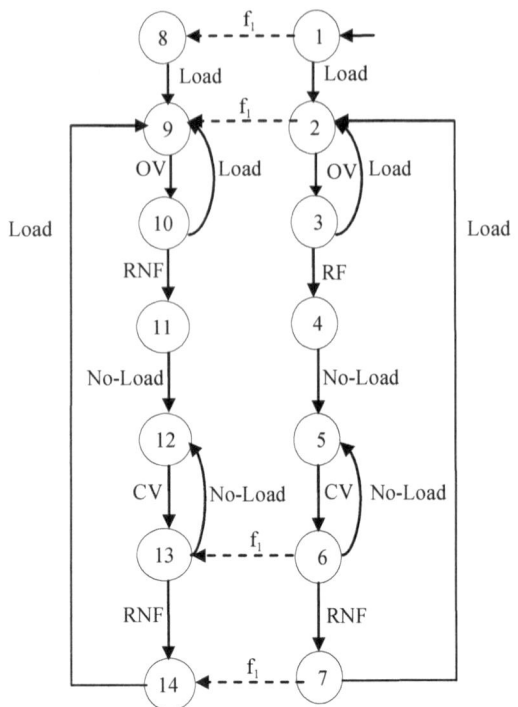

2.4 Supervision Pattern Approach

A supervision pattern Ω [22] is an automaton generating language $L(\Omega)$. The latter
is the set of trajectories (traces or event sequences) that one wants to recognize their
occurrence. The diagnoser in this case is expressed as a function defined on the set of
the event sequences that the system can generate. By observing the event sequences,
this function provides one of the following decisions: 'Yes', 'No' and '?'. When
the observed event sequence matches the pattern, the function provides the decision
'Yes' to indicate that it recognizes the pattern. It provides the decision 'No' when the
observed event sequence does not match the pattern. The decision '?' is issued when
the function cannot decide whether the event sequence matches the pattern or not.
This decision corresponds to the uncertainty or ambiguity case which we have seen
in the diagnoser approach. A state belonging to the subset, Q_F, of the states Q_Ω,
constituting the supervision pattern automaton, is reached when an event sequence
matching Ω has occurred.

The diagnoser recognizing the occurrence of a specified pattern is constructed as
follows:

• **Step 1**: Build supervision pattern Ω based on the use of global model G including
all the normal and faulty behaviors. Notice that G can be non-deterministic. Ω

Fig. 2.23 $G_\Omega = G x \Omega$
between model G (Fig. 2.22)
and supervision pattern Ω
(Fig. 2.19) for Example 2.11

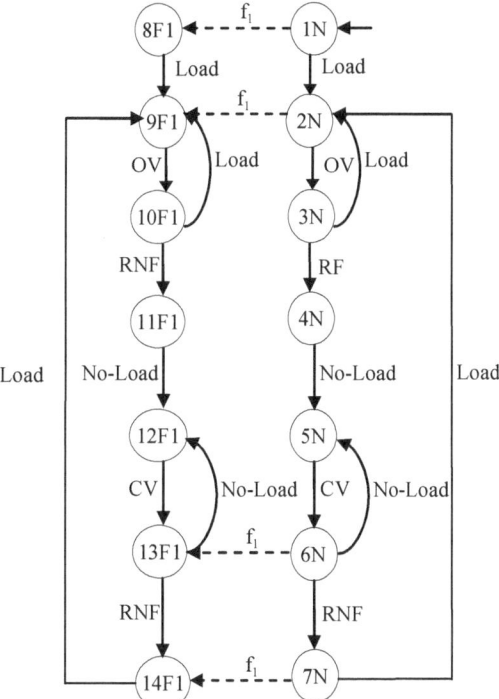

is represented by a particular Labeled Transition System (LTS) where each state contains a label to indicate if the specified behavior (supervision pattern Ω) has occurred or not. As we said before, any state belonging to Q_F is reached when an event sequence matching Ω has occurred. Thus, any event sequence u leading to reach a state of Q_F belongs to language $L_{Q_F}(\Omega)$. Ω is a deterministic in the sense that one event (observable or unobservable) can occur for each state and complete in the sense that all the events can occur for each state.

- **Step 2**: Achieve the product $G_\Omega = G x \Omega$ between the supervision pattern Ω and model G in order to label the states of G according to the supervision pattern Ω labels. The states of the corresponding LTS contain the labels indicating whether supervision pattern Ω has been recognized or not,

- **Step 3**: Construct the observable system (automaton) $OBS(G_\Omega)$ for G_Ω by eliminating from G_Ω all the transitions labeled by unobservable events (the event faults) and merging the states with their associated labels reached by the same transition labeled by the same observable events. If $OBS(G_\Omega)$ is deterministic, then it acts as a diagnoser since any of its state provides a label indicating whether the supervision pattern has occurred or not.

Definition 2.6 Ω certain state is a diagnoser state which indicates that the diagnoser recognizes with certainty the execution of supervision pattern Ω.

Ω uncertain state is a diagnoser state indicating that the diagnoser cannot recognize with certainty the execution of supervision pattern Ω.

Fig. 2.24 $OBS(G_\Omega)$ for the
example of Fig. 2.23

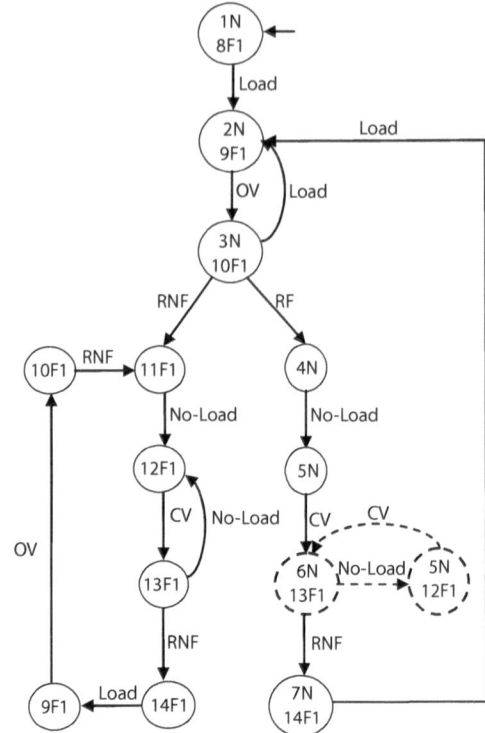

Fig. 2.25 Supervision pattern
Ω as Labeled Transition
System (LTS) for Example
2.12. Ω represents any faulty
behavior leading to both valve
stuck-close and pump failed
on failure modes

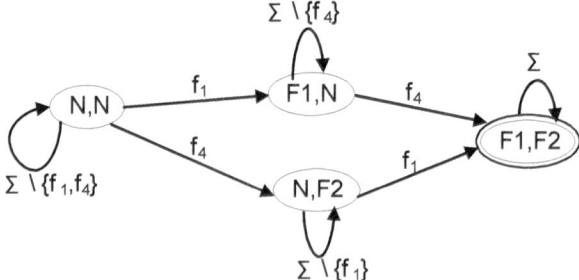

If Ω corresponds to the occurrence of a fault of type Γ_i, then certain state is
equivalent to F_i-uncertain state of the diagnoser approach (see Definition 2.3).

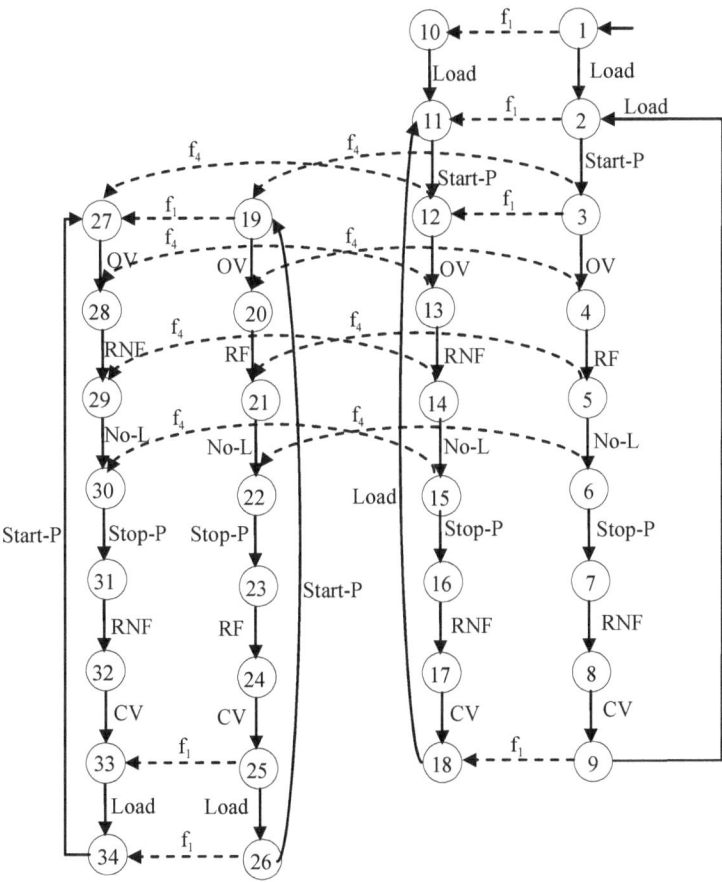

Fig. 2.26 Global model G for Example 2.12

Definition 2.7 Two conditions are required for a system to be Ω diagnosed [22]:

1. the execution of Ω must lead the diagnoser to visit Ω certain state,
2. the diagnoser must not contain any Ω-indeterminate cycle. As in the case of F_i-indeterminate cycle (Definition 2.4), Ω-indeterminate cycle in diagnoser D is a cycle composed exclusively of F_i-uncertain states.

Example 2.10 Let us take the example of the pump-valve system. Let us consider that the pump is always in its on-state. Let us consider that only the valve can fail in its stuck-close failure state (VSC) due to the occurrence of fault event f_1. Global model G of the system is depicted in Fig. 2.18. Supervision pattern Ω is the specified faulty behavior indicating that the valve is in its stuck-close failure state (*VSC*). Thus, it represents simply any behavior which can happen after the occurrence of fault f_1 as it is depicted in Fig. 2.19. We can notice that Ω is a deterministic and complete LTS. The product $G_\Omega = Gx\Omega$ between supervision pattern

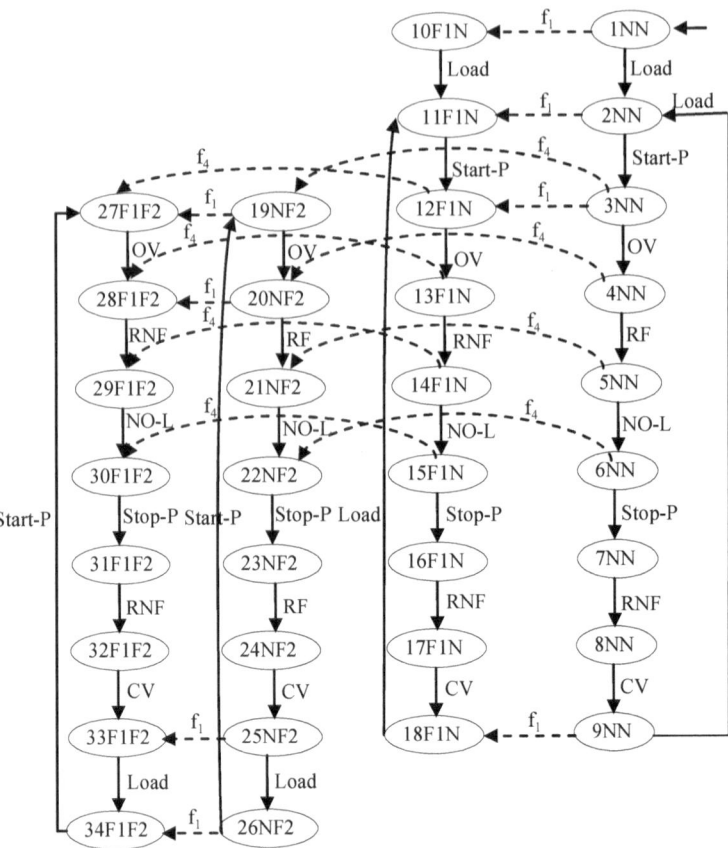

Fig. 2.27 $G_\Omega = Gx\Omega$ between the supervision pattern Ω (Fig. 2.25) and model G (Fig. 2.26) for Example 2.12

Ω and model G allows labelling the states of G according to Ω labels (F1 and N), as it is depicted in Fig. 2.20. Figure 2.21 shows the observable automaton $OBS(G_\Omega)$ for G_Ω. Since $OBS(G_\Omega)$ is deterministic, it corresponds to the diagnoser of pattern Ω representing the occurrence of the stuck close fault event.

Diagnoser state, Fig. 2.21, $q_D = (3\,N, 10F1)$ is Ω uncertain state (Definition 2.5) since it contains two labels one to indicate the non-execution of Ω and the other label to indicate its execution. Diagnoser state $q_D = 11F1$ is Ω certain state since it contains only the label indicating the execution of supervision pattern Ω. The diagnoser of Fig. 2.21 is Ω diagnosable (Definition 2.7) because the diagnoser can reach a Ω certain state (11F1) with a finite number of observable events and it does not contain any Ω-indeterminate cycle (Definition 2.6).

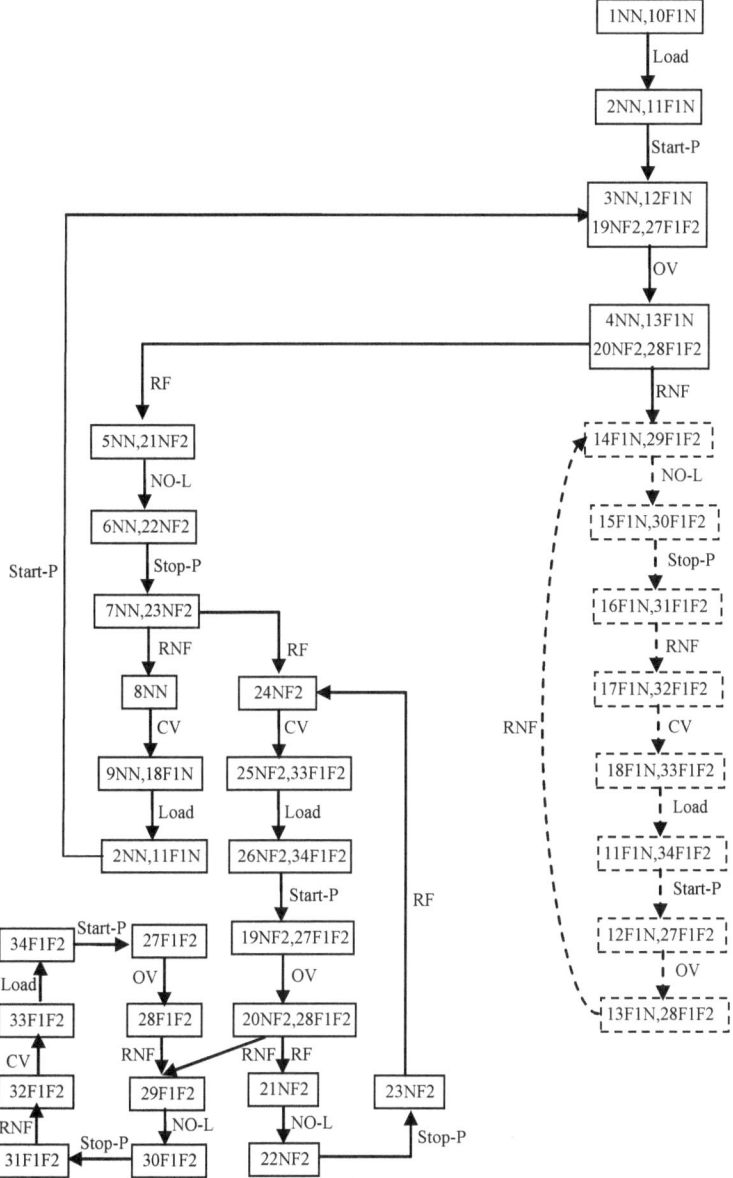

Fig. 2.28 $OBS(G_\Omega)$ of G_Ω (Fig. 2.27) for Example 2.12

Example 2.11 Let us consider Example 2.10 but with global model G depicted in Fig. 2.22. The LTS, obtained by the product $G_\Omega = Gx\Omega$ between supervision pattern Ω (Fig. 2.19) and model G (Fig. 2.22) in order to label the states of G according to Ω, is depicted in Fig. 2.23. Figure 2.24 shows the observable automaton $OBS(G_\Omega)$ for G_Ω. Since $OBS(G_\Omega)$ is deterministic, thus it corresponds to the diagnoser of supervision pattern Ω representing the occurrence of the stuck close fault event.

The system of Example 2.11 (see Fig. 2.24) is not Ω diagnosable (see Definition 2.7) because the diagnoser contains the Ω -indeterminate cycle ($< No$-$Load > < CV >)^*$ shown in dotted line in Fig. 2.24.

Example 2.12 Let us take the example of the pump-valve system. Let us consider that the valve can fail at the stuck-closed failure mode (occurrence of fault f_1 of type F_1), as it is depicted in Fig. 2.15a, and the pump can fail at failed-on failure mode (occurrence of fault f_4 of type F_2) as we can see in Fig. 2.15b. The controller model for this example is shown in Fig. 2.15c. We consider in this example that both the valve and the pump can fail; the multiple failures scenario can thus happen. Figure 2.25 shows supervision pattern Ω corresponding to the occurrence of both faults f_1 and f_4. Figure 2.26 shows global model G. The LTS obtained by the product $G_\Omega = Gx\Omega$, between supervision pattern Ω and model G in order to label the states of G according to Ω labels (F1,F2,N), is depicted in Fig. 2.27. Figure 2.28 shows the observable automaton $OBS(G_\Omega)$ for G_Ω. Since $OBS(G_\Omega)$ is deterministic, thus it corresponds to the diagnoser of supervision pattern Ω representing the occurrence of the valve stuck close and pump failed on fault events. We can notice the case of the occurrence of multiple failure modes (occurrence of both f_1 and f_4) represented by labels $F_1 F_2$.

This system is not Ω diagnosticable (see Fig. 2.28) because the diagnoser contains the $F_1 F_2$ indeterminate cycle ($< No$-$Load > < Stop$-$P > < CV > < RNF > < Load > < Start$-$P > < OV > < RNF >)^*$ showed in dotted line in Fig. 2.28. Indeed, the occurrence of the pump failed-on failure hides the occurrence of the valve stuck-close fault. As we have seen in Example 2.10, supervision pattern Ω can be diagnosed with certainty when the valve stuck-close fault occurs alone (see Fig. 2.21).

Chapter 3
Decentralized Diagnosis of Discrete Event Systems

3.1 Introduction

The main disadvantage of centralized diagnoser approaches is its space complexity. They require a centralized model in order to construct a centralized diagnoser. Both may become too large to be physically stored when a large-scale system as communication networks is under consideration. Moreover, if a centralized diagnoser exists physically, it suffers from the following problems [46]: (1) weak robustness because a partial malfunction of the centralized diagnoser may bring down the entire diagnosis task; (2) low maintainability due to the fact that any change in the system's structure may require a total redesign of the diagnoser, which can be very time-consuming and expensive and finally (3) the transmission of sensor data to the centralized diagnoser complicates the diagnosis reasoning because of the necessity to take into account the communication delay and the missing data or the errors in some sensor data during their transmission to the central point.

Therefore, the aim of using decentralized approaches is to overcome the space complexity and weak robustness of centralized approaches while at the same time preserving the diagnostic capability of a centralized diagnoser. In decentralized approaches, there are several local diagnosers, each of which receives observations from a specific area of the system and makes local diagnosis decision based on local observations. Ideally, there should be no communication between any pair of local diagnosers. However, the local or partial observation of the system may lead to high ambiguity of the final local diagnosis results. Therefore subject to physical restrictions, limited communication is permitted through a coordinator (see Fig. 1.7.b).

The decentralized diagnosis approaches are divided into two main categories: (1) the approaches using a global model in order to construct the local diagnosers and (2) the approaches using local models in order to construct the local diagnosers. Each of these two categories is divided into two main subcategories: the approaches considering a fault as the execution of an event and the approaches considering a fault as the violation of specifications. One typical approach of each of these categories will be developed in the next subsections.

M. Sayed-Mouchaweh, *Discrete Event Systems,* SpringerBriefs in Electrical and Computer Engineering, DOI 10.1007/978-1-4614-0031-8_3, © Author 2014

3.2   Co-Diagnosability Notion

The notion of co-diagnosability allows verifying whether a set of predefined faults can be diagnosed in decentralized manner using a set of local diagnosers. Each fault must be diagnosed by at least one local diagnoser by using its proper local observation of the system. Ideally, no communication is allowed between the local diagnosers. However, the co-diagnosability property is stronger than the diagnosability property (under the aggregated observation in the central point). If a system is co-diagnosable, then this means that it is diagnosable; while a diagnosable system does not ensure that it is co-diagnosable. In order to achieve a decentralized diagnosis equivalent to the one of a centralized diagnosis, a limited communication through a coordinator can be allowed between the different local diagnosers. This limited communication between local diagnosers is necessary in order to solve the ambiguity problem between local diagnosis decisions happened because of the partial observation of the system by each local diagnoser.

There are principally three notions to analyze the co-diagnosability property of system. These notions differ according to the interconnections between the system components. These notions will be studied in the following subtitles.

3.2.1   Local Diagnosability

A fault f occurring in a component or a subsystem i is local diagnosable [32] if and only if its occurrence is diagnosed using local diagnoser D_i associated to this component or subsystem i without the need to any communication with the other local diagnosers.

A component or subsystem i is local diagnosable if the occurrence of any fault f that can occur in component or subsystem i is diagnosable using only local diagnoser D_i associated to this component or subsystem i.

A system composed of n components or subsystems is local diagnosable if and only if each one of its components or subsystems is local diagnosable.

Example 3.1 Let us take the example of a system composed of two components modeled by the following two formal languages: $L_1 = \{u_{11} = o_1 o_2 c_1 c_2 (o_3)^*,$ $u_{12} = o_1 o_3 c_1 o_2 c_2 (o_2)^*, u_{13} = o_1 f_1 c_1 c_3 o_2 c_2 (o_3)^*\}$ and $L_2 = \{u_{21} = o_4 c_3 c_1 (o_2)^*,$ $u_{22} = o_4 f_2 c_2 c_1 o_5 (o_2)^*, u_{23} = o_4 c_2 o_5 o_2 c_2 (o_5)^*\}$. L_1 contains three event sequences: u_{11}, u_{12} and u_{13}. u_{11} and u_{12} are fault free event sequences while u_{13} is a faulty event sequence containing fault f_1. L_2 contains also three event sequences: u_{21}, u_{22} and u_{23}. u_{21} and u_{23} are fault free event sequences while u_{22} is f_2 fault event sequence. The set of observable events for these two components are, respectively: $\Sigma_{1o} = \{o_1, o_2, o_3\}$ and $\Sigma_{2o} = \{o_2, o_4, o_5\}$. Fault f_1 can occur only in component 1 while f_2 can occur only in component 2. Let P_1 and P_2 denote the projection functions of, respectively, components 1 and 2. P_1, respectively P_2, deletes from each event sequence the unobservable events by component 1, respectively component 2. Therefore, local diagnoser D_1 associated to component 1 can observe the following event sequences: $P_1(u_{11}) = o_1 o_2 (o_3)^*$, $P_1(u_{12}) = o_1 o_3 o_2 (o_2)^*$ and $P(u_{13}) = o_1 o_2 (o_3)^*$. We can see that D_1 cannot distinguish between fault free event sequence u_{11} and f_1 fault event sequence u_{13} since both have the same observable part: $P_1(u_{11}) = P_1(u_{13})$. Thus, component 1 is not local diagnosable. Local diagnoser D_2 associated to component 2 can observe the following event sequences: $P_2(u_{21}) = o_4 (o_2)^*$, $P_2(u_{22}) = o_4 o_5 (o_2)^*$ and $P_2(u_{23}) = o_4 o_5 o_2 (o_5)^*$. We can notice that f_2 fault event sequence u_{22} has an observable part, $o_4 o_5 (o_2)^*$, different from the one of the other two fault free event sequences u_{21} and u_{23}. Thus, component 2 is local diagnosable. The delay time to achieve the diagnosis of f_2 by D_2 is equal to the required time for the occurrence of the finite observable event sequence: $o_5 o_2 o_2$. However, the system composed by these two components is not local diagnosable since component 1 is not local diagnosable.

3.2.2 Independent Diagnosability

A fault f is independent diagnosable [44] if and only if its occurrence is diagnosable by one local diagnoser without the need to any communication with any other local diagnoser.

A component or a subsystem i is independent diagnosable if and only if the occurrence of any fault in this component or subsystem can be diagnosed independently by at least one local diagnoser.

A system composed of n components or subsystems is independent diagnosable if and only if each one of these components or subsystems is independent diagnosable.

The main difference between local diagnosable and independent diagnosable is that in the former a fault f occurring in component i must be diagnosed by diagnoser D_i associated to this component while in the latter f can be diagnosed by any other local diagnoser D_j where $j \neq i$. Therefore, the independent diagnosable notion requires the availability of the global model.

Example 3.2 Let us take a system composed of two components 1 and 2.: Let the formal language L generated by this system be: $L = \{u_{1f} = o_1 f o_3 c_2 o_2 (o_5)^*$, $u_2 = o_1 c_3 c_2 o_2 o_3 (o_5)^*$, $u_{3f} = o_2 c_2 f o_1 o_4 c_3 (o_4)^*$, $u_4 = o_2 o_4 c_1 o_4 c_3 o_1 (o_4)^* \}$. The sets of observable events by the two components 1 and 2 are, respectively: $\Sigma_{1o} = \{o_1, o_2, o_4, o_5\}$ and $\Sigma_{2o} = \{o_1, o_2, o_3, o_5\}$. There are two event sequences containing the fault f: $u_{1f} = o_1 f o_3 c_2 o_2 (o_5)^*$ and $u_{3f} = o_2 c_2 f o_1 o_4 c_3 (o_4)^*$. Let P_1 and P_2 denote the projection functions of, respectively, components 1 and 2. P_1, respectively P_2, deletes from each event sequence the unobservable events by component 1, respectively component 2. The observable parts of the event sequences of L according to each component are: $P_1(u_{1f}) = o_1 o_2 (o_5)^*$, $P_1(u_2) = o_1 o_2 (o_5)^*$, $P_1(u_{3f}) = o_2 o_1 o_4 (o_4)^*$ and $P_1(u_4) = o_2 o_4 o_4 o_1 (o_4)^*$ for the local diagnoser D_1 of component 1 and $P_2(u_1) = o_1 o_3 o_2 (o_5)^*$, $P_2(u_2) = o_1 o_2 o_3 (o_5)^*$, $P_2(u_{3f}) = o_2 o_1$ and

$P_2(u_4) = o_2 o_1$ for local diagnoser D_2 for component 2. We can notice that fault f is not diagnosed independently by D_1 and D_2. Indeed, D_1 cannot distinguish between fault free event sequence u_2 and fault event sequence u_{1f} since they have the same observable part, $P_1(u_2) = P_1(u_{1f})$, according to D_1. Same remark for D_2 since fault free event sequence u_4 has the same observable part, $P_2(u_4)$, as the one, $P_2(u_{3f})$, of fault event sequence u_{3f} according to D_2. Therefore, the system is not independent diagnosable.

Definition 3.1   If a system is local diagnosable then it is independent diagnosable. However, the contrary does not hold since the local diagnosability property is stronger than the independent diagnosability property.

3.2.3   Decentralized Diagnosability

A fault f is decentralized diagnosable [45] if and only if its occurrence is diagnosed with certainty and within a finite delay of time by the cooperation between local diagnosers.

A component or subsystem i is decentralized diagnosable if and only if each fault which can occur in this component or subsystem i is decentralized diagnosed.

A system composed of n components or subsystems is decentralized diagnosable if and only if each of its components or subsystems is decentralized diagnosable.

Example 3.3   Let us take the system of Example 3.2. We have seen in Example 3.2 that D_1 cannot distinguish between fault free event sequence u_2 and the faulty event sequence u_{1f} since they have the same observable part, $P_1(u_2) = P_1(u_{1f})$, according to D_1. Same remark for D_2 since the fault free event sequence u_4 has the same observable part, $P_2(u_4)$, as the one, $P_2(u_{3f})$, of faulty event sequence u_{3f} according to D_2. However, the system is co-diagnosable if a communication between these two local diagnosers through a simple coordinator based on a set of rules (conditional structure as we will see in 3.3.1) is allowed. Indeed, in the case of the occurrence of faulty event sequence u_{1f}, if D_2 does not declare normal operating conditions, then D_1 can declare the fault f since D_2 can distinguish between u_{1f} and u_2. Similarly, in the case of occurrence of faulty event sequence u_{3f}, if D_1 does not declare normal operating conditions, then D_2 can declare f since D_1 can distinguish between u_{3f} and u_4. The simple coordinator will use the following rule to issue its decision: 'F if neither D_1 nor D_2 declares N'. Therefore, this collaboration, or communication, between the two local diagnosers allows the system to be co-diagnosable. Thus, the system is decentralized diagnosable.

Definition 3.2   If a system is local diagnosable then it is independent and decentralized diagnosable. If a system is independent diagnosable then it is decentralized diagnosable. However, the contrary does not hold since the decentralized diagnosability property is weaker than local and independent diagnosability properties.

Example 3.4 Let us take the example of a system represented by formal language $L = \{u_{1f} = c_1 f o_1 o_3 o_4 c_2 o_2 o_3, u_2 = o_2 c_2 c_3 o_1 o_3 o_3 o_2, u_3 = o_3 o_1 c_2 o_4 o_2 o_3\}$. This system is observed by two sites. The sets of observable events for these two sites are: $\Sigma_{1o} = \{o_1, o_3\}$ and $\Sigma_{2o} = \{o_2, o_4\}$. Local diagnoser D_1, respectively D_2, is associated to site 1, respectively site 2. Both local diagnosers D_1 and D_2 cannot diagnose independently the occurrence of f because D_1 cannot distinguish fault event sequence u_{1f} from fault free event sequence u_2 and D_2 cannot differentiate between u_{1f} and u_3. Indeed, the observable part, $P_1(u_{1f}) = o_1 o_3 o_3$, of faulty event sequence u_{1f} according to D_1 is similar to the one, $P_1(u_2) = o_1 o_3 o_3$, of fault free event sequence u_2. Likewise the observable part, $P_2(u_{1f}) = o_4 o_2$, of faulty event sequence u_{1f} according to D_2 is similar to the one, $P_2(u_3) = o_4 o_2$, of fault free event sequence u_3. However, if a communication is allowed between both diagnosers D_1 and D_2 through a simple coordinator (see 3.3.1), the system becomes co-diagnosable. Indeed, D_1 can confirm the occurrence of fault free event sequence u_3 while D_2 can confirm the occurrence of fault free event sequence u_2. If both do not confirm the occurrence of a fault free event sequence, then the coordinator decision will be the occurrence of fault event sequence u_{1f}. The simple coordinator will use the following rule to issue its decision:
  'F if neither D_1 nor D_2 declares N'.

3.3 Decentralized Diagnosis Based on a Global Model

Two main approaches for decentralized diagnosis using a global model will be studied: the approaches based on the use of a conditional structure and approaches using a complex coordinator.

3.3.1 Decentralized Approaches with Conditional Structure

Typical example of decentralized approaches with conditional structure is the one developed in [49]. In order to explain this approach, let us take the example of Fig. 3.1. This example presents a system composed of pump P and three valves V_1, V_2 and V_3. Valves V_1 and V_2 are connected to pump P through valve V_3. The output of valve V_1, respectively V_2, is equipped by flow sensor s_1, respectively s_2. s_1, respectively s_2, provides the event RF_1, respectively RF_2, when there is a flow at the output of valve V_1, respectively V_2, and RNF_1, respectively RNF_2, when there is no flow. Two faults f_1 valve V_3 is stuck-off and f_2 (pump is failed off) are considered for this example. The occurrence of these two faults is indicated, respectively, by the fault labels F_1 and F_2.

The set of observable events for this system is:

$$\Sigma_o = \{Start - P, Stop - P, OV_1, CV_1, OV_2, CV_2, OV_3, CV_3, RF_1,$$
$$RNF_1, RF_2, RNF_2\}$$

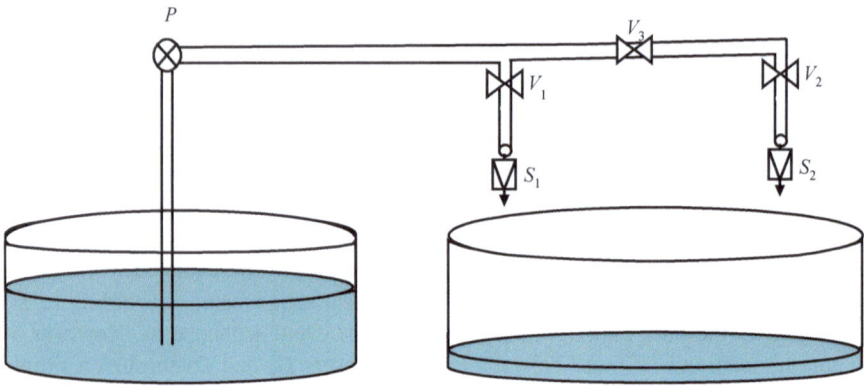

Fig. 3.1 Two tanks system example

where:

Start-P: Controllable event to start pump P,
Stop-P: Controllable event to stop pump P,
OV_1: Controllable event to open valve V_1,
CV_1: Controllable event to close valve V_1,
OV_2: Controllable event to open valve V_2,
CV_2: Controllable event to close valve V_2,
OV_3: Controllable event to open valve V_3,
CV_3: Controllable event to close valve V_3.

In the initial state of the system, pump P is in its off state; all the valves are closed. Global model G of this system is depicted in Fig. 3.2. We suppose that flow Q_p of pump P is bigger than both flow Q_{v1} of valve V_1 and flow Q_{v2} of valve V_2: $Q_p > Q_{v1} + Q_{v2}$. In this model, f_1 can happen whenever valve V_3 is in its closed state and f_2 can occur only when pump P is in its off state as we have seen before in Fig. 2.3 and 2.4.

We suppose that the system is divided into two sites; the set of observable events for sites 1 and 2 are, respectively:

$$\Sigma_{1o} = \{Start - P, Stop - P, OV_1, CV_1, RF_1, RNF_1\},$$
$$\Sigma_{2o} = \{OV_2, CV_2, OV_3, CV_3, RF_2, RNF_2\}.$$

Let us start by constructing local diagnosers D_1 and D_2 for sites 1 and 2 based on the same reasoning followed in chapter 2 but using, respectively, the set of observable events Σ_{1o} and Σ_{2o}. Here, no need to integrate the sensors readings to each event since we consider their reading as an event. The two local diagnosers are constructed based on the use of global model G of Fig. 3.2 and are depicted, respectively, in Fig. 3.3 and 3.4.

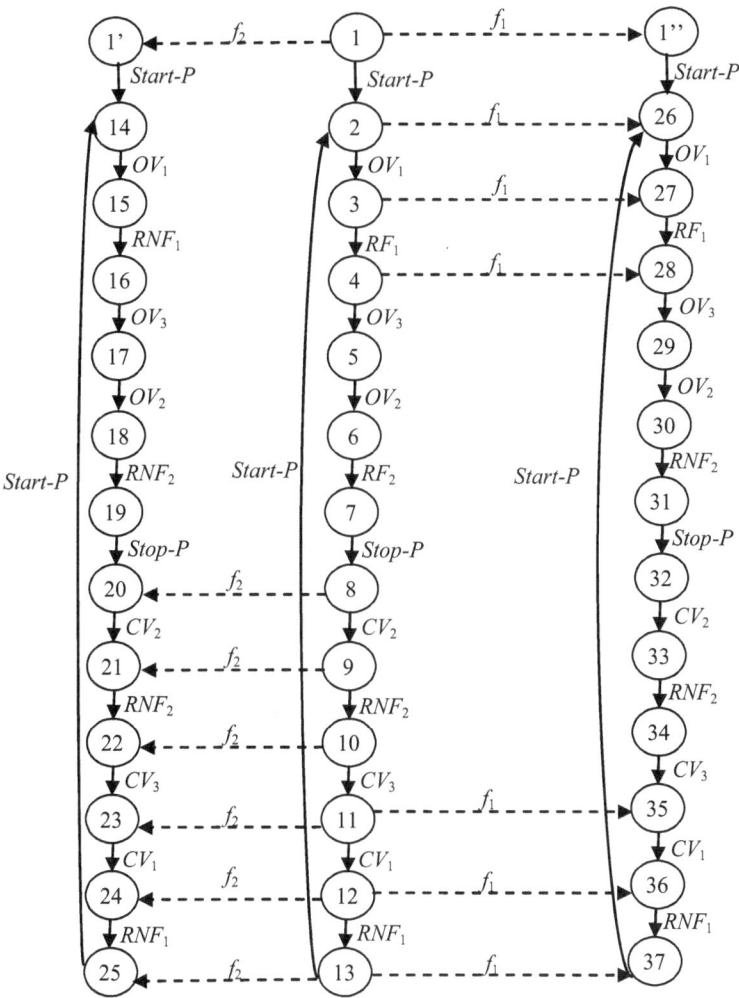

Fig. 3.2 Global model *G* for the example of Fig. 3.1

By observing Fig. 3.3 and 3.4, we can notice that both local diagnosers D_1 and D_2 cannot diagnose with certainty the occurrence of fault f_1 of type F_1 indicating that valve V_3 is stuck-off. Indeed, D_1 can diagnose with certainty the occurrence of f_2 of type F_2 while it cannot distinguish between the normal functioning, N, and the occurrence of fault f_1 of type F_1. D_2 can ensure the normal functioning but it cannot distinguish between the occurrence of f_1 and f_2. This confusion is due to the partial observation of these two local diagnosers. Indeed, three event sequences are generated by the system: fault-free event sequence u_N, f_1 fault event sequences u_{F1} indicating the occurrence of fault f_1 and f_2 fault event sequences u_{F2} indicating the

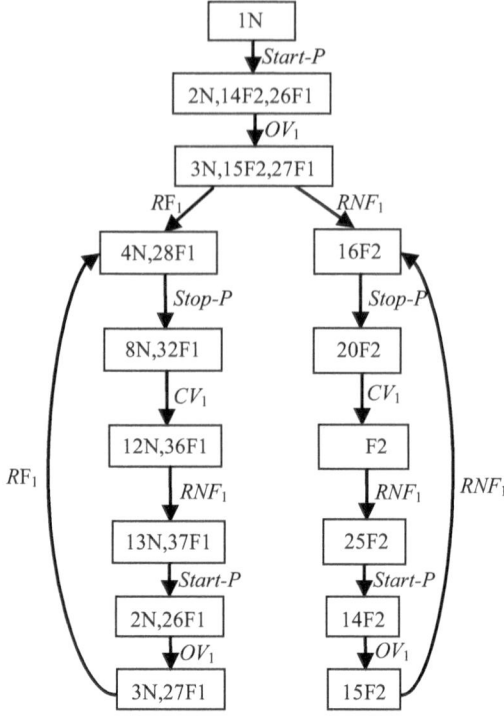

Fig. 3.3 Local diagnoser D_1 for global model G of Fig. 3.2

occurrence of fault f_2. The observable parts of these event sequences after removing unobservable events (faulty events f_1 and f_2) are:

$$P(u_N) = <Start\text{-}P><OV_1><RF_1><OV_3><OV_2><RF_2>$$
$$<Stop\text{-}P><CV_2><RNF_2><CV_3><CV_1><RNF_1>$$
$$P(u_{F1}) = <Start\text{-}P><OV_1><RF_1><OV_3><OV_2><RNF_2>$$
$$<Stop\text{-}P><CV_2><RNF_2><CV_3><CV_1><RNF_1>$$
$$P(u_{F2}) = <Start\text{-}P><OV_1><RF_1><OV_3><OV_2><RNF_2>$$
$$<Stop\text{-}P><CV_2><RNF_2><CV_3><CV_1><RNF_1>$$

A centralized diagnoser can distinguish between each of these event sequences and thus can infer with certainty the occurrence of f_1 or f_2 since $P(u_N) \neq P(u_{F1}) \neq P(u_{F2})$.

These event sequences are observed by D_1 as follows:

$$P_1(u_N) = <Start\text{-}P><OV_1><RF_1> \quad <Stop\text{-}P><CV_1><RNF_1>$$
$$P_1(u_{F1}) = <Start\text{-}P><OV_1><RF_1> \quad <Stop\text{-}P><CV_1><RNF_1>$$
$$P_1(u_{F2}) = <Start\text{-}P><OV_1><RNF_1> \quad <Stop\text{-}P><CV_1><RNF_1>$$

We can see that D_1 can distinguish between u_N and u_{F2} and thus can infer the occurrence of f_2 with certainty. However, it cannot distinguish between u_N and u_{F1}.

Fig. 3.4 Local diagnoser D_2
for global model G of Fig. 3.2

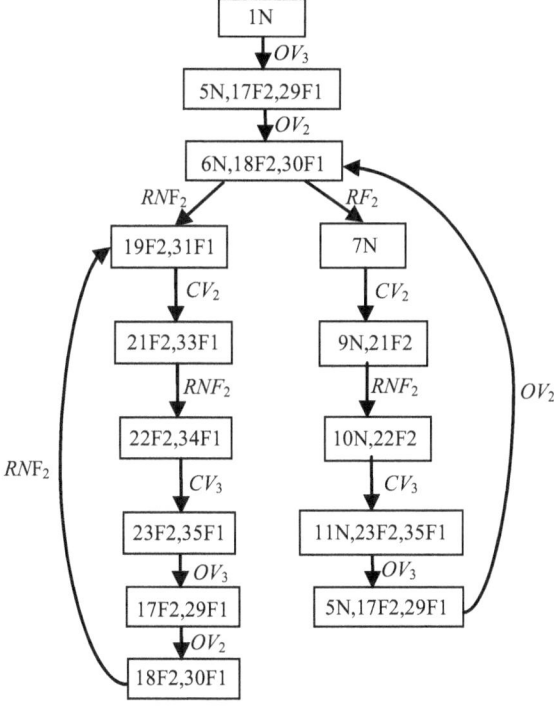

These event sequences are observed by D_2 as follows:

$$P_2(u_N) = <OV_3><OV_2><RF_2><CV_2><RNF_2><CV_3>$$

$$P_2(u_{F1}) = <OV_3><OV_2><RNF_2><CV_2><RNF_2><CV_3>$$

$$P_2(u_{F2}) = <OV_3><OV_2><RNF_2><CV_2><RNF_2><CV_3>$$

We can see that D_2 can distinguish between u_N and both u_{F1} and u_{F2}. Thus, D_2 can infer with certainty the non-occurrence of f_1 and f_2. However, it cannot distinguish between u_{F1} and u_{F2}.

This decentralized diagnosis structure is not equivalent to a centralized diagnosis structure since neither D_1 nor D_2 can diagnose with certainty the occurrence of f_1. A decentralized diagnosis structure equivalent to a centralized one can be obtained by using a set of rules of the form 'D_1 declares F_i if D_2 does not declare N'. Indeed, if we take the following event sequence corresponding to the occurrence of f_1:

$$<Start\text{-}P><OV_1><RF_1><OV_3><OV_2><RNF_2>$$

As we have said before, a global diagnoser can infer with certainty the occurrence of f_1. This event sequence will lead D_1 to the state (4N,28F1) and D_2 to the state (19F2,31F1) as we can see, respectively, in Fig. 3.3a and 3.3b. Both diagnosers cannot infer the occurrence of f_1. However, since D_2 can infer with certainty the

Table 3.1 Decision rules for the decentralized diagnosis structure of Fig. 3.3 and 3.4

Local diagnoser D1	Local diagnoser D2	Global decision
N	N	N
F2	Nothing	F2
F1 if D2 does not declare N	Nothing	F1
F1 if D2 does not declare N	N	N
Nothing	Nothing	Nothing

non-occurrence of f_1 and f_2, thus if D_2 does not declare N with certainty then D_1 can declare F_1. In other words, f_1 can be inferred by using the following rule when both diagnosers D_1 and D_2 say nothing, i.e., a local diagnoser declares nothing when it cannot confirm the occurrence or the non-occurrence of a fault:

'D_1 declares F_1 if D_2 does not declare N'

Table 3.1 shows the required rules for the decentralized diagnosis with conditional structure of Fig. 3.3 and 3.4. The decentralized diagnosis structure becomes conditional because of the use of these rules which enable this decentralized structure to achieve a diagnosis performance equivalent to the one of a centralized structure.

Local diagnoser D_1 (Fig. 3.3) can diagnose with certainty and within a finite delay time the occurrence of the fault f_2 (pump failed-off) which can happen only in its subsystem. Thus, subsystem 1 is local diagnosable. Local diagnoser D_2 (Fig. 3.4) cannot diagnose with certainty the occurrence of fault f_1 (valve V_3 stuck-off) occurring only in its associated subsystem. Thus, subsystem 2 is not local diagnosable. Consequently, the system of Fig. 3.1 is not local diagnosable since its subsystem 2 is not local diagnosable. This is expected result since the system requires a simple coordinator based on the use of a set of rules, Table 3.1, in order to be co-diagnosable.

Since subsystem 1 is local diagnosable then it is independent diagnosis (see Definition 3.1). Indeed, fault f_2 can be diagnosed by D_1 without any communication with the other local diagnoser D_2. However, the system of Fig. 3.1 is not independent diagnosable since its subsystem 2 is not independent diagnosable. Indeed, Local diagnoser D_2 cannot diagnose alone and without any communication the occurrence of fault f_1 (valve V_3 stuck-off) occurring only in its associated subsystem.

The system becomes co-diagnosable if a communication between both diagnosers D_1 and D_2 is allowed throughout a simple coordinator based on the use of a set of rules as we can see in Table 1.3. Thus, the system of Fig. 3.1 is decentralized diagnosable.

3.3.2   Coordinated Structure

When a decentralized diagnosis with conditional structure is unable to achieve a diagnosis performance equivalent to the one of a centralized diagnosis, a decentralized diagnosis with a complex coordinator is required. Typical example of decentralized diagnosis using coordinated structure is the approach proposed in [12]. In this approach, the system is observed through two sites. Each site comprises two modules:

local observation module and extended local diagnoser. The local observation module P_i observes a specific part of the system through a mask or projection function. Only a subset Σ_{io} of the set of observable events Σ_o is observed by site i where $\Sigma_{io} \subset \Sigma_o$. Extended local diagnoser D_i^e collects information about the system operating conditions (state) through the local observation mask P_i and calculates the fault label(s) using this partial observation about the system. The global diagnosis decision is calculated using a coordinator. The latter merges the extended local diagnosers' decisions using a set of decision rules. The latter require a set of registers to stock the local diagnosis decisions as well as the estimate of the previous states reached due to the occurrence of unobservable events by one of local extended diagnosers. The coordinated diagnosis structure ensures a diagnosis performance equivalent to the one of a centralized diagnoser.

The following hypothesis must hold [12]:

- language $L(G)$ generated by model G is live in order to ensure that there are no deadlocks;
- G has no cycles of unobservable events with respect to sites 1 and 2 in order to ensure that the system will not stay infinitely without generating observable events which are used to infer the fault occurrence;
- the faults cannot be all diagnosed by one extended local diagnoser in order to eliminate the trivial case where the system remains diagnosable in a centralized way by one local diagnoser;
- the messages sent from a local site are received correctly and in the order that they sent to the coordinator;
- the coordinator does not have a copy of the system model. It has only a limited memory and limited processing capabilities.

Let us take the following example of Fig. 3.5 in order to explain how coordinated decentralized diagnosis structure proposed in [12] works. This example shows a system composed of pump P, three valves V_1, V_2 and V_3 and three tanks T_1, T_2 and T_3. Pump P provides a flow Q_p to valves V_1 and V_2. Valves V_1 and V_2 have different sections and thus different output flows Q_1 and Q_2. They aim at filling the two connected tanks T_1 and T_2 of equal surfaces. Tank T_1, respectively T_2, is equipped by a level sensor s_1, respectively s_2, in order to generate the event RL_1, respectively RL_2, when the liquid level in the tank 1, respectively tank 2, is equal to L_1, respectively L_2. Tank T_2 is equipped with drain valve V_3 in order to evacuate tanks T_1 and T_2 into tank T_3. The global model for this example is depicted in Fig. 3.6. For the sake of simplicity, we suppose that:

- only valve V_1 can fail in the stuck-off failure mode;
- tank 1 level L_1 is equal to tank 2 level L_2;
- the sections of V_1, V_2 and the connected pipe are chosen to ensure that $Q_1 > Q_2 + Q_3$;
- after reaching L_1 and L_2, the controller evacuates the content of both tanks T_1 and T_2 using valve V_3 before starting a new cycle. For the sake of simplicity, this evacuation is not shown in global model G of Fig. 3.5.

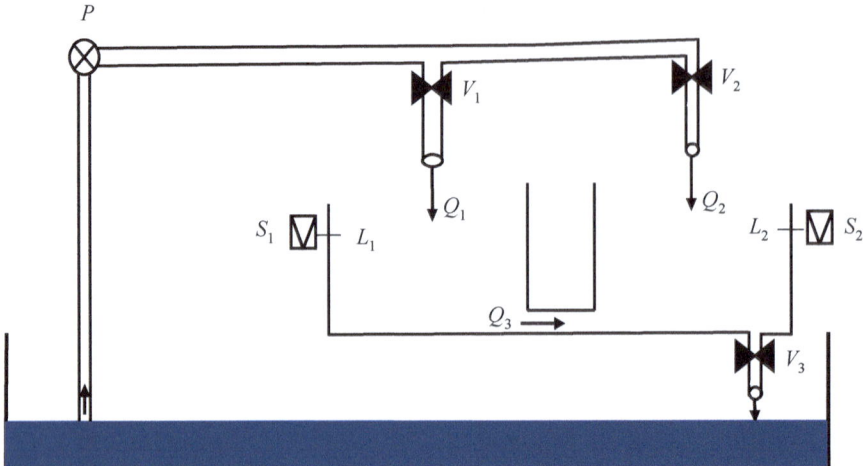

Fig. 3.5 Example of two connected tanks

Fig. 3.6 Global model G for
the example of Fig. 3.5

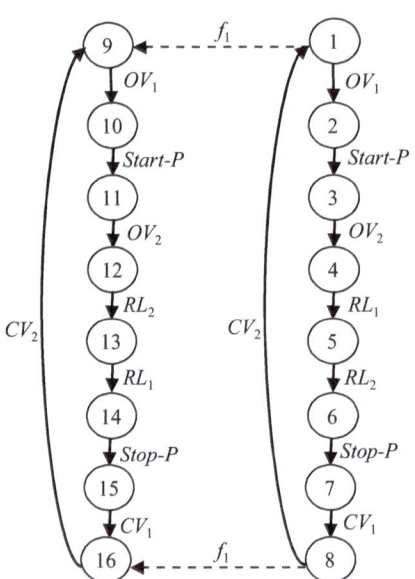

We suppose that the system is divided into two sites:

- Site 1 observes the set of observable events $\Sigma_{1o} = \{Start - P, Stop - P, OV_1, CV_1, RL_1\}$;
- Site 2 observes the set of observable events $\Sigma_{2o} = \{Start - P, Stop - P, OV_2, CV_2, RL_2\}$.

Fig. 3.7 Extended local diagnosers for the example of Fig. 3.6

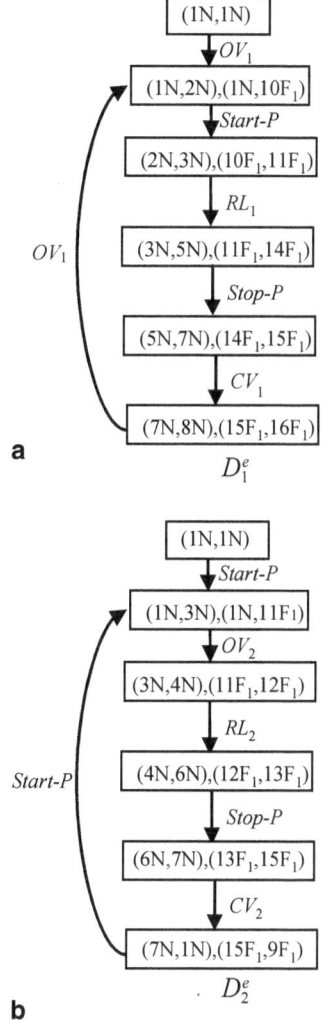

The two local extended diagnosers D_1^e and D_2^e are shown, respectively, in Fig. 3.7a and 3.7b. We can notice that:

- a global diagnoser can distinguish between the normal event sequence u_N and the faulty event sequence u_F since their observable parts $P(u_N)$ and $P(u_F)$ are different:

$$P(u_N) = <OV_1><Start\text{-}P><OV_2><RL_1><RL_2><Stop\text{-}P><CV_1><CV_2>$$

$$P(u_F) = <OV_1><Start\text{-}P><OV_2><RL_2><RL_1><Stop\text{-}P><CV_1><CV_2>$$

- both local diagnosers cannot diagnose with certainty the occurrence of fault f_1 due to their partial observation of the system. Indeed, local diagnoser D_1, respectively

D_2, cannot distinguish between the normal and faulty event sequences, $P_1(u_N) = P_1(u_F)$, respectively $P_2(u_N) = P_2(u_F)$, using their own observation:

$$P_1(u_N) = <OV_1><Start\text{-}P><RL_1><Stop\text{-}P><CV_1>$$
$$P_1(u_F) = <OV_1><Start\text{-}P><RL_1><Stop\text{-}P><CV_1>$$
$$P_2(u_N) = <Start\text{-}P><OV_2><RL_2><Stop\text{-}P><CV_2>$$
$$P_2(u_F) = <Start\text{-}P><OV_2><RL_2><Stop\text{-}P><CV_2>$$

• this ambiguity cannot be solved using a conditional structure since neither of the local diagnosers can ensure the normal operating conditions.

In order to solve this ambiguity, a complex coordinator needs to be constructed. This coordinator is based on the use of 5 registers R_1, R_2, R_3, R_4 and C as well as a set of decision rules handling the local diagnosis decisions as well as the estimated global system states. R_1 and R_2 register, respectively, the actual states of extended local diagnosers D_1^e and D_2^e. R_3 registers the unobservable reach of D_1^e due to the occurrence of unobservable events by D_1^e. R_4 registers the unobservable reach of D_2^e due to the occurrence of unobservable events by D_2^e. C registers the global diagnosis decision.

Example 3.5 In order to understand the functioning of these registers, let us take the example of Fig. 3.5. At the system initial state, both extended diagnosers are in their initial states $q^1 = q^2 = (1N,1N)$ (see Fig. 3.7a and 3.7b). Thus, the contents of R_1, R_2 and C are equal to $(1N,1N)$. The first event that can occur starting from the system initial state, 1N, is event OV_1 (see Fig. 3.6). This event is observable by D_1^e (see Fig. 3.7a) and unobservable by D_2^e (see Fig. 3.7b). Since starting from the initial state of D_1^e, there is no unobservable event that can occur, thus the unobservable reach $UR(q^1)$ of D_1^e will be only the current state $q^1 = (1N,1N)$ of D_1^e. Therefore, the content of R_3 is equal to $(1N,1N)$. The occurrence of event OV_1, observable by D_1^e, will move D_1^e to $q^1 = \{(1N,2N), (1N,10F1)\}$ as we can see in Fig. 3.7a. Therefore, the content of R_1 is equal to $\{(1N,2N),(1N,10F1)\}$. The occurrence of unobservable event OV_1 by D_2^e will lead the system to the state 2 with the label N or to the state 10 with the fault label F_1 (see Fig. 3.6). Thus, the unobservable reach $UR(q^2)$ of D_2^e is equal to $\{(1N,2N),(1N,10F1)\}$. We add also to this unreachable reach the current state $q^2 = (1N,1N)$ of D_2^e and register them in R_4. Therefore, R_4 is equal to $\{(1N,1N),(1N,2N),(1N,10F1)\}$. Since event OV_1 is not observable by D_2^e, the content of R_2 will not change and will be equal to $(1N,1N)$.

Example 3.6 Let us take now the second event that can occur in the system. This event is *Start-P* which is observable by D_1^e and D_2^e. Thus, the unobservable reach registered in R_3 and R_4 will not change. The occurrence of this event will move the state of D_1^e from $\{(1N,2N),(1N,10F1)\}$ to $\{(2N,3N),(10F1,11F1)\}$ as we can see in Fig. 3.7a. Therefore, the content of R_1 is equal to $\{(2N,3N),(10F1,11F1)\}$. Likewise, the occurrence of *Start-P* will move the state of D_2^e from $(1N,1N)$ to $\{(1N,3N),(1N,11F1)\}$ as we can see in Fig. 3.7b. Therefore, the content of R_2 is equal to $\{(1N,3N),(1N,11F1)\}$.

In order to calculate the new global coordinator state in response to the occurrence of new observable event, β, preceded by observable event α by one or both local extended diagnosers, one of the following decision rules must be applied. The application of one of these decision rules will update the content of register C in order to include the new global system states associated to the corresponding fault labels. These decision rules are:

$$DR_1 : C_{New} = (R_1 \cap_C^L R_4) \cap_C C_{Old} : \alpha, \beta \in \sum_{1o} \backslash \sum_{2o} \tag{3.1}$$

$$DR_2 : C_{New} = (R_1 \cap_C^R R_4) \cap_C C_{Old} : \beta \in \sum_{1o} \backslash \sum_{2o}, \alpha \in \sum_{2o} \backslash \sum_{1o} \tag{3.2}$$

$$DR_3 : C_{New} = (R_2 \cap_C^L R_3) \cap_C C_{Old} : \alpha, \beta \in \sum_{2o} \backslash \sum_{1o} \tag{3.3}$$

$$DR_4 : C_{New} = (R_2 \cap_C^R R_3) \cap_C C_{Old} : \beta \in \sum_{2o} \backslash \sum_{1o}, \alpha \in \sum_{1o} \backslash \sum_{2o} \tag{3.4}$$

$$DR_5 : C_{New} = (R_1 \cap_C^L R_2) \cap_C C_{Old} : \beta \in \sum_{1o} \cap \sum_{2o}, \alpha \in \sum_{1o} \backslash \sum_{2o} \tag{3.5}$$

$$DR_6 : C_{New} = (R_1 \cap_C^R R_2) \cap_C C_{Old} : \beta \in \sum_{1o} \cap \sum_{2o}, \alpha \in \sum_{2o} \backslash \sum_{1o} \tag{3.6}$$

"$\backslash$" is the set difference operator.

$\cap_C^L$ is the intersection left between two registers and it is computed by finding the common final states in both registers and then by appending their original states in the register in the left of this intersection. $\cap_C^R$ is computed as $\cap_C^L$ but the original states are appended using the register in the right of this intersection.

Example 3.7 In order to understand the functioning of the intersections left and right, let us calculate the intersection left between the contents of the following two registers:

$$\{(4N, 6N), (12F1, 13F1)\} \cap_C^L \{(3N, 4N), (11F1, 13F1), (2N, 3N), (10F1, 11F1)\}$$

The only common final state between the registers in the left and in the right of the intersection left $\cap_C^L$ is 13F1. The original state, predecessor, of 13F1 in the register in the left of the intersection left $\cap_C^L$ is 12F1. Thus, the result of this intersection is equal to $\{(12F1,13F1)\}$.

For the case of intersection right, let us take the same example: and calculate the intersection right between the contents of the same registers:

$$\{(4N, 6N), (12F1, 13F1)\} \cap_C^R \{(3N, 4N), (11F1, 13F1), (2N, 3N), (10F1, 11F1)\}$$

The original state, predecessor, of 13F1 in the register in the right of the intersection right $\cap_C^R$ is 11F1. Thus, the result of this intersection $\cap_C^R$ is equal to $\{(11F1,13F1)\}$.

The coordinator intersection $\cap_C$ is computed as follows. We look for the final states of the register in the right of coordinator intersection $\cap_C$ which are in common with the original states of the register in the left of $\cap_C$. Then, the final result will be these original states with their corresponding final states.

Example 3.8 In order to understand the functioning of coordinator intersection, let us calculate the coordinator intersection between the contents of the following registers

$$\{(1N, 2N), (1N, 10F1)\} \cap_C (1N, 1N)$$

The final state of the register in the right of $\cap_C$ is 1N. The latter is the original state of both final states 2N and 10F1. Thus, the result of this intersection is these two final states with their original states: $\{(1N, 2N), (1N, 10F1)\}$.

Example 3.9 Let us see now how the coordinator can update its global state using one of the decision rules defined by Eq. (3.1) to (3.6). Event OV_1 belongs to $\Sigma_{1o} \backslash \Sigma_{2o}$, thus DR_1, defined by (3.1), will be applied in order to update the register C content. As we have seen in Example 3.5, the contents of registers R_1, R_4 and C_{old} are, respectively: $\{(1N,2N),(1N,10F1)\}$, $\{(1N,1N),(1N,2N),(1N,10F1)\}$ and $(1N,1N)$. Therefore, the new content of C is achieved as follows:

$$C_{new} = (\{(1N, 2N), (1N, 10F1)\} \cap_C^R \{(1N, 1N), (1N, 2N), (1N, 10F1)\}) \cap_C (1N, 1N)$$

$$C_{new} = (\{(1N, 2N), (1N, 10F1)\}) \cap_C (1N, 1N) = \{(1N, 2N), (1N, 10F1)\}.$$

Example 3.10 Let us see now how the coordinator can solve the decision ambiguity due to the partial observation of local diagnosers. Let us take the faulty event sequence $u_F = <OV_1><Start\text{-}P><OV_2><RL_2>$ and let us see how the coordinator can find the right global state with its fault label after the occurrence of each observable event by one or both of local diagnosers. After the occurrence of observable event RL_2, the registers have the following contents:

$$C_{Old} = \{(3N, 4N), (11F1, 12F1)\}$$
$$R_1 = q^1 = \{(2N, 3N), (10F1, 11F1)\}$$
$$R_2 = q^2 = \{(4N, 6N), (12F1, 13F1)\}$$
$$R_3 = UR(q^1) = \{(3N, 4N), (11F1, 13F1), (2N, 3N), (10F1, 11F1)\}$$
$$R4 = UR(q^2) = \{(1N, 1N), (1N, 2N), (1N, 10F1)\}$$

Since this event is observable by the second site as well as the previous event, OV_2, then DR_3, defined by (3.3), will be applied as follows:

$$C_{New} = (R_2 \cap_C^L R_3) \cap_C C_{Old} : \alpha = OV_2, \beta = RL_2 \in \sum_{o2} \backslash \sum_{o1}$$
$$R_2 \cap_C^L R_3 = \{(12F1, 13F1)\};$$

$$C_{New} = \{(12F1, 13F1)\} \cap_C C_{Old} = \{(3N, 4N), (11F1, 12F1)\} = \{(12F1, 13F1)\}.$$

By analyzing the extended local diagnosers (Fig. 3.7) for the example of Fig. 3.5, we can notice that the system is not local diagnosable. Indeed, fault f_1, indicating the stuck-off failure of valve V_1, occurring in subsystem 1 cannot be diagnosed with certainty by extended local diagnoser D_1^e associated to this subsystem as we can see in Fig. 3.7a. The system is also not independent diagnosable. Indeed, the fault f_1 (stuck-off failure of valve V_1) cannot be diagnosed with certainty neither by D_1^e nor by D_2^e without any communication between them, as we can see in Fig. 3.7. However, the system becomes decentralized diagnosable when a communication between both extended local diagnosers D_1^e and D_2^e is allowed throughout a complex coordinator based on the use of a set of decision rules and a state estimation, as we have seen in Example 3.10.

3.3.3   Approaches Based on the Use of Specifications

In these approaches, a failure is considered as the violation of the desired behavior described by a set of specifications. A specification is a sub automaton of the system model. The set of specifications generates specification language K which is a part of system model language L. The violation of each specification indicates the occurrence of a fault. The fault diagnosis in these approaches aims at detecting a violation of any of the specifications and at determining which specification has been violated. Example of this category of approaches is the one developed in [36]. In order to explain this approach, let us take the example representing pump P and valve V. Figure 3.8 represents global model G of these two components. G represents all the feasible sequences of commands that these components can execute. Figure 3.9 shows the desired behavior or the specification that the system must satisfy. Figure 3.10 shows the augmented specification after adding fault state F indicating the violation of this specification. The other states have the fault free label, N, since they are corresponding to the desired behavior. This violation is due to the occurrence of a controller fault. The set of events is $\Sigma = \{f, Start - P, Stop - P, OV, CV\}$ where f is unobservable event occurring in the case of controller error. The set of observable events is thus $\Sigma_o = \{Start - P, Stop - P, OV, CV\}$. We suppose that the system is observed by two sites 1 and 2. Site 1, respectively site 2, observes the system through projection function P_1, respectively P_2. The set of observable events for site 1, respectively site 2, is $\Sigma_{1o} = \{Start - P, Stop - P\}$, respectively $\Sigma_{2o} = \{OV, CV\}$. Figure 3.11 shows the product combination $G \parallel \overline{R}$ between global model G (Fig. 3.8) and augmented specification $\overline{R}$ (Fig. 3.10). $G \parallel \overline{R}$ is an automaton. Each state of the latter is of the form $((q_{1G}, q_{1\overline{R}}), ..., (q_{jG}, q_{j\overline{R}}))$. $G \parallel \overline{R}$ represents the global diagnoser, because it associates to the states of model G the fault label, F, indicating the specification violation or the fault free label, N, indicating the specification satisfaction. Figure 3.12a and 3.12b show, respectively, local diagnosers $D_1 = P_1(G \parallel \overline{R})$ and $D_2 = P_2(G \parallel \overline{R})$ for, respectively, sites 1 and 2.

Example 3.11 State (1F) in the diagnoser $(D = G \parallel \overline{R})$ in Fig. 3.11 is F-certain state since the model state, state 1, is associated to fault label F indicating the violation of

Fig. 3.8 Global model G for the example of pump P and valve V

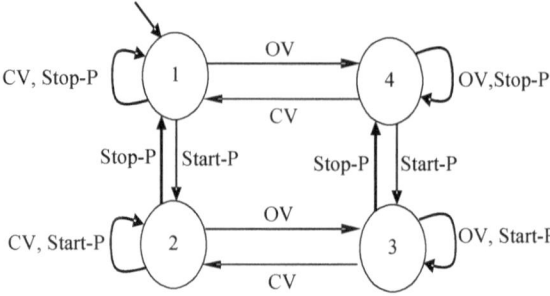

Fig. 3.9 Specification (desired behavior) R for the example of pump P and valve V

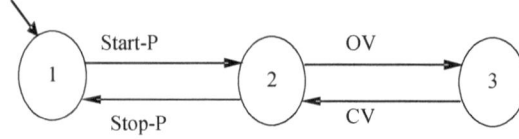

Fig. 3.10 Augmented specification $\bar{R}$ for the example of pump P and valve V

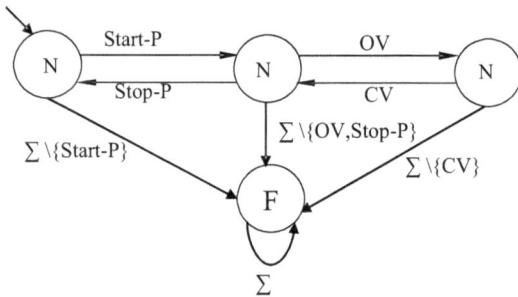

the specification depicted in Fig. 3.9. State (2N) of the same diagnoser in Fig. 3.11 is fault free or N-certain state because fault label F is not associated to any of its model G states. State (2N,3F) of local diagnoser $D_1 = P1(G\|\bar{R})$ of Fig. 3.12a is F-uncertain state because fault free label N is associated to model G state 2 and fault label F is associated to state 3 of model G (see Definition 2.3).

The system of Fig. 3.8 is not co-diagnosable since both diagnosers D_1 and D_2 cannot infer the occurrence of the controller fault when command event sequence $u_F = <OV><Start\text{-}P>$ is issued by the controller. Indeed, both diagnosers are in uncertain states, (2N, 3 F) for D_1 and (3N, 4 F) for D_2, as we can see in Fig. 3.12a and 3.12b. A centralized diagnoser, $D = G \| \bar{R}$, can diagnose this fault with certainty by reaching the certain states (4F) after the occurrence of OV and then the certain state (3F) after the occurrence of $Start\text{-}P$ (see Fig. 3.11). In order to obtain a decentralized diagnosis structure equivalent to a centralized one, a coordinator is required. This coordinator use one of the decision rules, $(DR_1, \ldots, DR_6)$ defined by Eq. (3.1) to (3.6), to calculate the global state of the system based on the local diagnosis decisions provided by D_1 and D_2.

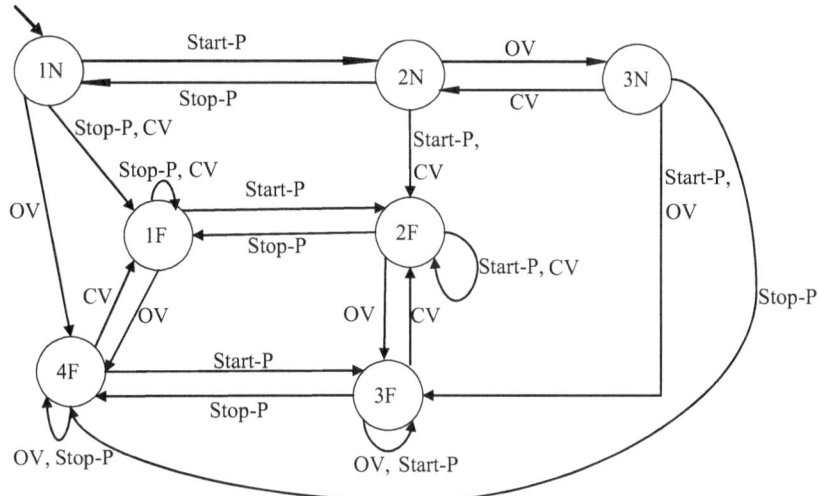

Fig. 3.11 Product combination $G \| \overline{R}$ between global model G (Fig. 3.8) and augmented specification $\overline{R}$ (Fig. 3.10)

Example 3.12 This example shows how the coordinator infers the occurrence of controller fault when fault event sequence $u_F = OV$ is issued by the controller. Before the occurrence of OV, the registers of the coordinator contain the following information: $C_{old} = (1N,1N)$, $R_1 = (1N,1N)$, $R_2 = (1N,1N)$. The occurrence of OV is observable by D_2 and unobservable by D_1. Therefore, the decision rule DR_3, defined by (3.3), will be applied:

$$C_{New} = (R_2 \cap_C^L R_3) \cap_C C_{Old}$$

The unobservable reach of D_1 can be calculated as follows. D_1 is in model G state 1 (see Fig. 3.11). At this state, three unobservable event sequences can occur: $(<OV>), (<CV>)$ and $(<OV><CV>)$. These unobservable event sequences will lead D1 from, respectively, 1N to 4F, 1N to 1F and 4F to 1F. Note for the last case that, although state 1N is the predecessor of 1F through $(<OV><CV>)$, the immediate predecessor of 1F, which is 4F, must be included in the tuple of unreachable reach. Therefore the content of R3 is equal to $\{(1N,4F),(1N,1F),(4F,1F)\}$. The contents of the other registers are:

$$R_2 = \{(1N, 3N), (1N, 4F)\}$$
$$R_2 \cap_C^L R_3 = (1N, 4F)$$

The update of the content of register C indicating the new global model state with its corresponding label after the occurrence of OV is:

$$C_{New} = (1N, 4F) \cap_c (1N, 1N) = (1N, 4F)$$

Therefore, the system is decentralized diagnosable using a complex coordinator.

Fig. 3.12 Diagnosers D_1 and D_2 for, respectively, sites 1 and 2 for the example of pump P and valve V of Fig. 3.7

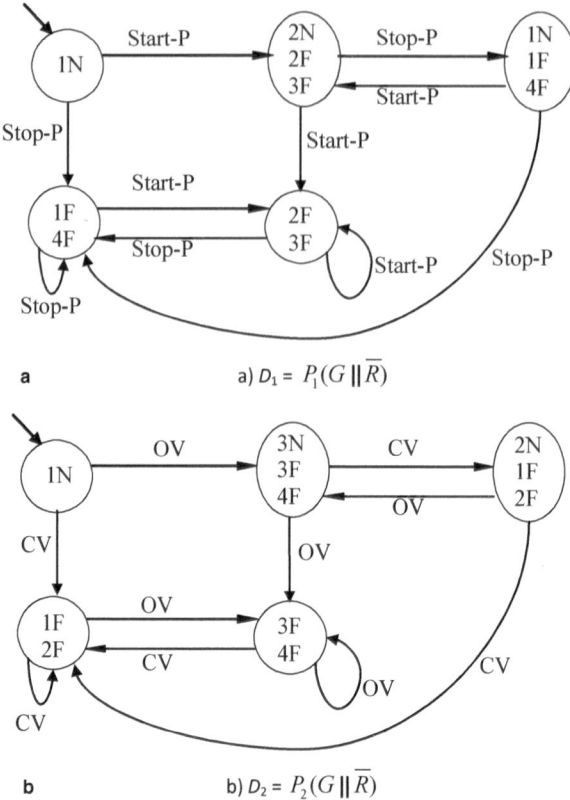

a)

a) $D_1 = P_1(G \| \overline{R})$

b)

b) $D_2 = P_2(G \| \overline{R})$

The system is not independent diagnosable because neither local diagnoser D_1, Fig. 3.12a, nor local diagnoser D_2, Fig. 3.12b, can diagnose with certainty and without any communication between them the violation of the specification of Fig. 3.9.

3.4 Decentralized Diagnosis Without the Use of a Global Model

3.4.1 Approaches Based on the Use of Specifications

Let us take the same example of pump P and valve V used previously for Sect. 3.3.3. In the latter, local diagnosers D_1 and D_2 for sites 1 and 2 were constructed using global model G. In this section, local diagnosers D_1 and D_2 will be constructed using local models G_1 and G_2 of Fig. 3.13. G_1 represents the local model for the pump and G_2 is the local model of the valve. Figure 3.14a and 3.14b show local diagnosers D_1 and D_2 constructed using local models G_1 and G_2. We can notice that these

Fig. 3.13 Local models G_1 and G_2 for the example of pump P and valve V

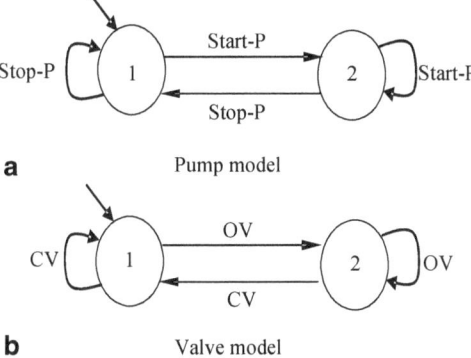

a   Pump model

b   Valve model

diagnosers are similar to the ones of Fig. 3.12a and 3.12b. However, the difference between the local diagnosers of Fig. 3.12 constructed using a global model and the local diagnosers of Fig. 3.14 constructed using a local model is that in the first case, the diagnoser states contain the system global states while in the second case, they contain the system local states. Thus, when the local diagnosers are constructed using local models, the system becomes not co-diagnosable using a decentralized structure because a coordinator cannot be constructed. Indeed, the coordinator requires the use of a global model in order to estimate the global states of the system. This is a main inconvenient of the approaches based on the use of specifications without a global model. However, if the system is local diagnosable, then this approach becomes useful since the local diagnosers are constructed using local models and the decentralized diagnosis structure is equivalent to the centralized diagnosis structure.

We must indicate that if the generated language K by global specification model R is $\{\Sigma_i\}$ separable [52], then $D_1 = G_1 \parallel \overline{R_1}$ and $D_2 = G_2 \parallel \overline{R_2}$ where $R_1 = P_1(R)$ and $R_2 = P_2(R)$. A specification language K is said to be $\{\Sigma_i, i \in \{1,2,...,n\}\}$ separable, if and only if $R = R_1 \parallel R_2 \parallel ... \parallel R_n$. For our example of the pump and the valve, we can notice easily that K is not $\{\Sigma_i\}$ separable since $R \neq R_1 \parallel R_2$.

3.4.2   Approaches Based on Synchronous Composition of Local Diagnosers

Typical application example of this category of diagnosis approaches is the monitoring of hierarchical telecommunication networks [31–33]. The latter are composed of one supervision center (SC) and a set of technical centers (TC). Each TC has a set of data switches (SW) routing data through the network. SC is in charge of receiving alarms of each SW through its corresponding TC. We must note that the model size of such system is potentially too large. A simplified model of a communication network, as the one studied in [31], contains $2^{10} \times 4^{300}$ states. Therefore, the diagnosis of such system using a global model is unrealistic and unfeasible.

Fig. 3.14 Diagnosers D_1 and D_2 for, respectively, sites 1 and 2 based on the use of local models G_1 and G_2 (Fig. 3.12) for the example of pump P and valve V of Fig. 3.8

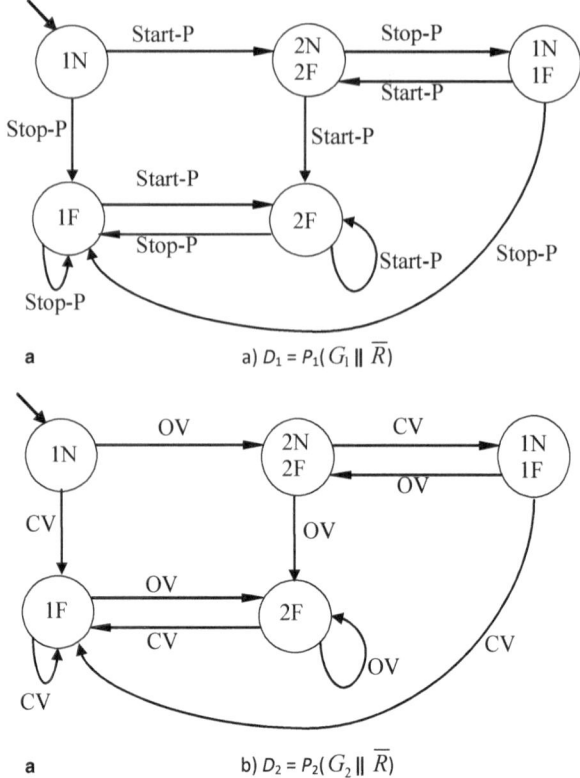

Faults in a telecommunication networks can be related to:

- communication part as a loss of messages or an error of message reception or emission;
- service quality as a too long response time;
- software treatment as insufficient memory (saturation of the buffers) or program code error;
- physical or hardware equipment as a cable cut or processor problem;
- exogenous events from the system environments as a problem of ventilation or a fire.

Alarms are observable events sent to the SC in order to be analyzed with the objective to determine the responsible component of this alarm. The events indicating a lock, unlock or rebooting of a SW or a TC are examples of alarms received by the SC. The events indicating the standby or recover of a component (SW or TC) are examples of unobservable events. An alarm is not enough to establish a diagnosis, i.e. localizing the responsible component causing this alarm. Indeed, the occurrence of a primary failure on a component may cause the occurrence of several secondary failures on other components. The secondary failures, caused by the propagation of primary failures, can also interfere. This failure propagation and interference may cause the reception of a huge number of observations, alarms, by the SC which complicates significantly the diagnosis task [33]. Another monitoring problem is related to the masking phenomenon. Indeed, the occurrence in the past of other primary or secondary failures may mask the occurrence of a part of observable events normally generated when a failure occurs. This will increase the number of failures that can occur without observable consequences and therefore the number of possible explanations, failure candidates, for a given set of observations [33].

A typical approach belonging to this category of methods is the one developed in [31]. In this approach, the fault diagnosis is achieved by analyzing the observed sequences of alarms, observable events, received by the SC and emitted by each SW through TC. Two kinds of failures can occur in each component: exogenous and internal failures. The exogenous failures are caused by the system environments while internal failures are caused by the failure propagation. The cut of physical connection between two switches because of an external action or the degradation of a component service quality over time due to its wearing are examples of exogenous failures. Internal failures are the direct or indirect consequences of the occurrence of at least one exogenous failure. Placing switches on mode standby is an example of an internal failure as a direct consequence of the exogenous failure represented by the cut of the switches' physical connection. The occurrence of new internal failures as the consequence of the propagation of internal failures towards other components is the indirect consequence of exogenous failures. Both kinds of failures are considered as unobservable events. A component reacts to both kinds of failures and emits observable events outside the system and unobservable events to other components. The events emitted to other components affect their behavior. Therefore, this approach decomposes the system into a set of interconnected components throughout the internal events in order to model the failure propagation.

In order to explain this approach, let us take the example of Fig. 3.15 based on an example extracted from [31]. The system is composed of two interconnected components, e.g., data switches. The sets of observable events by components 1 and 2 are respectively: $\Sigma_{1o} = \{o_{11}, o_{12}\}$ and $\Sigma_{2o} = \{o_{22}, o_{21}\}$. Failures of types F_1 and F_2 can occur in component 1 while failures of types F_3 and F_4 can occur in component 2. Only the failure of type F_1 can be propagated from component 1 to component 2 through the emission of internal event i_{12}. Similarly, failures of types F_3 and F_4 can be propagated from component 2 to component 1 through the emission of internal event i_{21}. As it has been indicated before, all the failures and internal events are unobservable events. Figure 3.15 shows the local models G_1 and G_2 for components 1 and 2. The decentralized diagnosis using this approach is calculated using the following steps:

- **Step 1:** *Construct local diagnosers D_1 and D_2 construction for components 1 and 2 based on the use of their local models G_1 and G_2. D_1 is built as follows (see Fig. 3.16). Each state of D_1 comprises all local model G_1 states that can be reached by the same observable event. In other words, a local diagnoser state contains all the states reached by an event sequence having as observable projection the same*

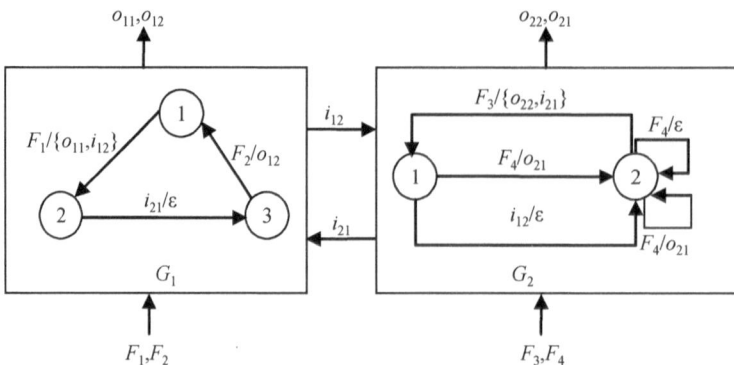

Fig. 3.15 Example of a system composed of two interconnected components representing a communication network. G_1 is the local model of the first component and G_2 is the local model of the second component.

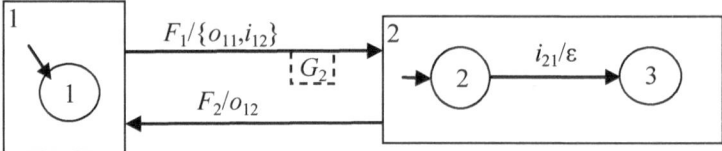

Fig. 3.16 Local diagnoser D_1 of local model G_1 for the example of Fig. 3.15

observable events. As an example, the occurrence of observable event o_{11} will change the local diagnoser D_1 state from the local model G_1 state 1 to its state 2 then to its state 3 due to the occurrence of unobservable event i_{21} as we can see in Fig. 3.16. Thus, local diagnoser D_1 state 2 will contain G_1 state 2 linked to G_1 state 3 by the transition labeled by unobservable event i_{21} as we can see in Fig. 3.16. Local diagnoser D_1 will return from its state 2 to its initial state 1 after the occurrence of observable event o_{12}. Each transition of local diagnoser D_1 may be labelled by the synchronization label 'G_2' if the occurrence of the corresponding observable event generates an internal event emitted to local model G_2. As an example, the occurrence of observable event o_{12} in Fig. 3.16 generates internal event i_{12}. The latter changes the state of local model G_2 of component 2. Therefore, synchronization label is necessary to synchronize the state change of local diagnosers D_1 and D_2 of both components since the generated internal event will change at the same time the state of G_2. Same reasoning can be followed for the construction of local diagnoser D_2 for local model G_2 of component 2 as we can see in Fig. 3.17.

- **Step 2:** *Compute local diagnosis decisions after the occurrence of an observable event.* As an example, the occurrence of the observable event sequence ($<o_{11}><o_{22}>$) will change the local diagnoser D_1 state from 1 to 2 after the occurrence of o_{11} and the local diagnoser D_2 state from 1 to 2 after the occurrence of

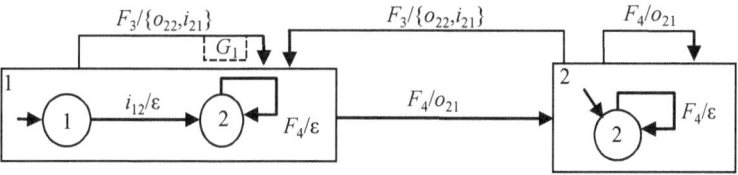

Fig. 3.17 Local diagnoser D_2 of local model G_2 for the example of Fig. 3.15

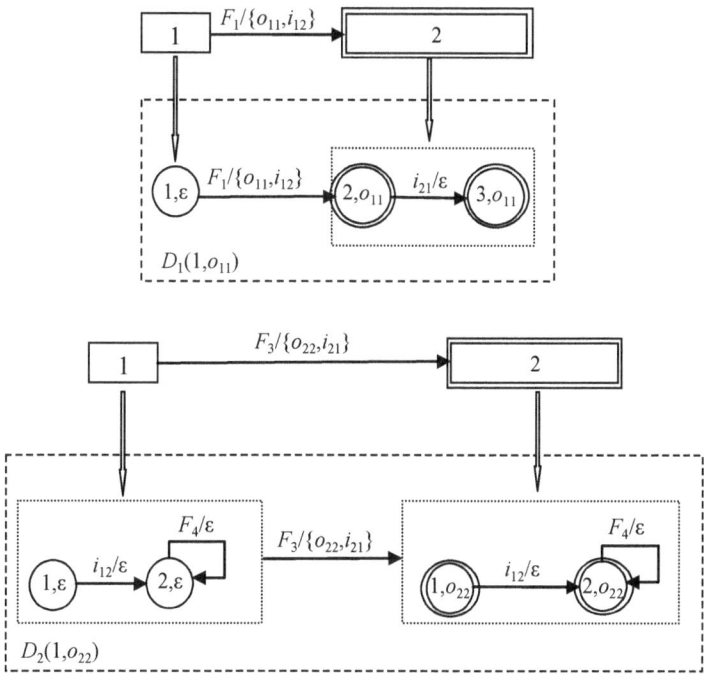

Fig. 3.18 Computing local diagnosis decisions $D_1(1,o_{11})$ and $D_2(1,o_{22})$ in response to the occurrence of event sequence ($<o_{11}><o_{22}>$) for the example of Fig. 3.15

o_{22} as we can see in Fig. 3.18. Therefore, the next state of D_1 after the occurrence of o_{11}, $D_1(1,o_{11})$, is equal to the local diagnoser D_1 state 2 and the next state of D_2 after the occurrence of o_{22}, $D_2(1,o_{22})$, is equal to the local diagnoser D_2 state 2. We must mention here that both diagnosers are not yet synchronized. Thus, the combination of local model states equivalent to the global model state is not yet computed.

- **Step 3:** *Calculate the global diagnosis decision using a coordinator.* In this the step, the local diagnosis decisions will be synchronized in order to compute the combination of local models' states equivalent to the global model state. This combination of local models' states represents the explanation to the observed event or alarms sequence. This explanation is then used to determine the sequence of

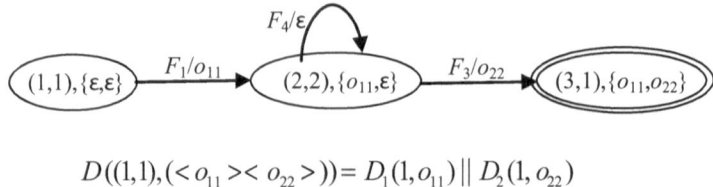

$$D((1,1),(< o_{11} >< o_{22} >))= D_1(1,o_{11}) \| D_2(1,o_{22})$$

Fig. 3.19 Computing the global diagnosis decision D in response to the occurrence of event sequence $(<o_{11}><o_{22}>)$ for the example of Fig. 3.15. Global decision D is obtained by the synchronization of both $D_1(1,o_{11})$ with $D_2(1,o_{22})$ of Fig. 3.18

fault labels generating this alarm sequence. The sequence of fault labels explaining the alarm sequence can be infinite as we will see later through the example of Fig. 3.15. The synchronization is achieved using a coordinator. The latter synchronizes the local diagnosis decisions between the interacting components through internal events using the standard synchronous composition operator $\|$. The coordinator considers two local models as interacting if and only if each of respective local diagnosers claims that it interacts with the other one. The synchronization is achieved in parallel and through different stages. The goal of this parallelization and decomposition into several stages is to accelerate and to minimize the combination process efforts of the different local diagnosis decisions. Indeed, the standard synchronous composition operator $\|$ is commutative and associative. Thus, the composition between several local diagnosers decisions can be achieved in many different ways. To optimize the composition, one way is to compose the local diagnosis decisions between direct interacting components in order to eliminate in the first stage all incompatible behaviors (states and transitions) and to obtain smaller models to be combined in next stages. The information about the components interacting directly can be extracted from the synchronous label attached to local diagnoser transitions as we can see in Fig. 3.16 and 3.17. Let us take the example of Fig. 3.18 and let us compute the global diagnosis decision $D((1,1),(<o_{11}><o_{22}>))= D_1(1,o_{11})\|D_2(1,o_{22})(1,1)$ in D represents the combination of local models states and $(<o_{11}><o_{22}>)$ denotes the observable event sequence. Since the synchronization label attached to the transition corresponding to the occurrence of observable event o_{11} in D_1 (see Fig. 3.16) indicates the interaction between components 1 and 2 through internal event i_{12} and the transition corresponding to the occurrence of o_{22} in D_2 (see Fig. 3.17) claims the interaction between both components through internal event i_{21}, then the coordinator considers that both components are interacting directly and their local diagnosis decisions must be synchronized. The global diagnosis decision is obtained by achieving the synchronization operation between both local diagnosers. The synchronization of local diagnosis decision $D_1(1,o_{11})$ in Fig. 3.16 with local diagnosis decision $D_2(1,o_{22})$ in Fig. 3.17 will lead to obtain the global diagnosis decision of Fig. 3.19. The occurrence of event sequence $(<o_{11}><o_{22}>)$ is explained by the fault sequence $F_1(F_4)^* F_3$. If both local diagnosers D_1 and D_2 were not synchronized, D_1 will explain this event sequence by fault sequence F_1 and D_2 by fault sequence F_3 which leads to obtain fault sequence $F_1 F_3$. The latter does not correspond to the right one, $F_1(F_4)^* F_3$, explaining the occurrence of event or alarm sequence ($<o_{11}><o_{22}>$).

The example of Fig. 3.15 is not local diagnosable since the local diagnosers, Fig. 3.16 and 3.17, need to be synchronized in order to diagnose the faults of types F_1, F_2, F_3 and F_4. Thus, the system requires a communication through a coordinator in order to be co-diagnosable. This is an expected result since faults can be propagated from one component to another one.

By following the same reasoning, the example of Fig. 3.15 is not independent diagnosable since the local diagnosers, Fig. 3.16 and 3.17, need to be synchronized through a coordinator in order to diagnose faults of types F_1, F_2, F_3 and F_4. However, the system becomes co-diagnosable when a communication through a coordinator is allowed. This communication is the synchronization between local diagnosers in order to take into account the fault propagation from one component to another one. Thus, the system of Fig. 3.15 is decentralized diagnosable.

Chapter 4
Conclusion and Discussion

Generally, the methods for fault diagnosis of discrete event systems are divided into three categories: centralized [22, 23, 27, 39], decentralized [4, 12, 31, 35, 36, 41, 49, 52] and distributed [10, 16, 18, 24, 46] methods. The centralized methods construct one global diagnosis module based on the use of a global model of the system to be monitored. They can be divided, at their turn, into two main categories: diagnoser [39] and supervision pattern approaches [22]. In diagnoser approaches, a fault is considered as the execution of an event and the model G represents the normal and faulty behaviors of the system. The fault diagnosis is based on the observation of observable event sequences after the fault occurrence. In supervision pattern approaches, faults are represented as the execution of specified faulty behaviors (supervision patterns). A faulty behavior is modeled as a set of partial observable trajectories (traces or event sequences) that one wants to recognize their occurrence. The diagnosis of the occurrence of f is achieved by the matching between the real behavior of the system and the compiled faulty behaviors.

In both diagnoser and supervision pattern approaches, the diagnoser is constructed off-line and then it is used on-line to diagnose the occurrence of one of the predefined faults. Diagnoser approach has the advantage to construct the diagnoser in automatic manner thanks to the use of the synchronous composition operator between the different system components models; while the supervision pattern requires the global model of the different components constituting the whole system. However, both approaches suffer from the following drawbacks:

- the faults to be diagnosed must be defined in advance and must be included in the components models or in the global model;
- the size of the diagnoser is exponential with respect to the number of system components and discrete variables;
- the diagnoser does not integrate the time of events occurrences. Thus, the drift-like faults cannot be diagnosed by these approaches. Indeed, drift-like faults are characterized by too late or too early occurrence of events. These two characteristics cannot be represented by diagnoser or supervision pattern approaches.

The advantage of supervision pattern approach according to the diagnoser approach is the fact that the former can be used to diagnose intermittent faults while the latter

Fig. 4.1 Supervision pattern
Ω as a Labeled Transition
System (LTS) for the
occurrence of an intermittent
fault f. r is the repair event

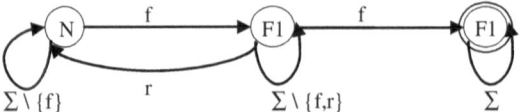

cannot diagnose these faults. Indeed, when more than one failure of the same type,
say F_i, occurs along an event sequence u of the generated language L by the model
G with or without a repair, diagnoser approach does not require that each of these
occurrences be detected. It is enough to conclude, within finite number of events
after the occurrence of the first failure, that along u, a failure of type F_i happened. In
supervision pattern approach, intermittent faults can be represented as a supervision
pattern indicating the occurrence of fault f of Type F_i twice (or k times) without
repair [22] as we can see in Fig. 4.1.

For any centralized diagnoser, the diagnosability property must be analyzed. The
latter can be defined as the capacity of a diagnoser to infer the occurrence of a fault
as well as its fault partition based on the observation of the sequences of observable
events. The diagnosability property can be verified using a diagnosability notion.
The latter is based on the fact that a system is diagnosable if and only if any pair of
faulty/non-faulty behaviors can be distinguished by their projections to observable
behaviors.

Centralized diagnosis approaches are not suitable for large scale systems as the
telecommunication networks. Indeed, in the latter, the global model can contain a
huge number of states. As an example, for telecommunication networks, as the one
studied in [31, 32, 33], the number of states of the global model is of the order of
$2^{10} \times 4^{300}$ [31]. Therefore, constructing the global model is physically unfeasible.
Moreover, if a centralized diagnoser exists physically, it suffers from the following
problems [46]: (1) weak robustness because a partial malfunction of the centralized
diagnoser may bring down the entire diagnosis task; (2) low maintainability due to
the fact that any change of the system's structure may require a total redesign of
the diagnoser, which can be very time-consuming and expensive. Therefore, the aim
of using decentralized approaches is to overcome the space complexity and weak
robustness of centralized approaches while at the same time preserving the diagnostic capability of a centralized diagnoser. In these approaches, there are several
local diagnosers, each of which receives observations from a specific area of the system and makes local diagnosis decision based on local observations. Ideally, there
should be no communication between any pair of local diagnosers. However, the
local or partial observation of the system may lead to high ambiguity of the final
local diagnosis results. Therefore, very limited communication is permitted through
a coordinator. However, the main disadvantage of these approaches is the need of a
global model to construct the local diagnosers as well as the coordinator. An alterna-
tive to this disadvantage is the distributed diagnosis approaches. In these approaches,
the system is divided into subsystems. Each subsystem knows only its own part of
the global model. A local diagnoser is associated to each subsystem in order to perform diagnosis locally. This diagnosis computation is based on the local model and

Fig. 4.2 Links between local, independent and decentralized diagnosability notions

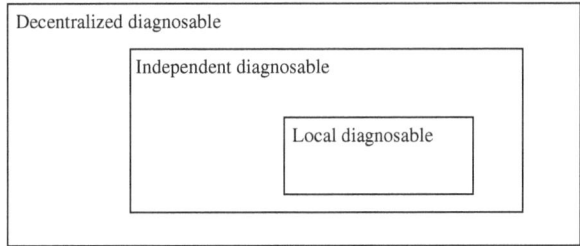

the information communicated directly to it by the other local diagnosers through a communication protocol. The information exchanged among local diagnosers is used by them to update their own information. This information update compensates the local diagnosers partial observation. However, a communication protocol must be defined in order to achieve consistency among local diagnosers. If there are no constraints imposed on the interactions among local models (subsystems), as hierarchical or tree structure, the communication protocol definition will lead to high time consumption and complexity state space.

Consequently, the challenge of decentralized/distributed diagnosis methods is to perform local diagnosis equivalent to the global one using a scalable communication protocol (with respect to the number of component modules) or without the need to a global model.

The notion of co-diagnosability allows verifying whether a set of predefined faults can be diagnosed in decentralized manner using a set of local diagnosers. Each fault must be diagnosed by at least one local diagnoser by using its proper local observation of the system. Ideally, no communication is allowed between the local diagnosers. However, the co-diagnosability property is stronger than the diagnosability property. If a system is co-diagnosable, then this means that it is diagnosable; while a diagnosable system does not ensure that it is co-diagnosable. In order to achieve a decentralized diagnosis equivalent to the one of a centralized diagnosis, a limited communication through a coordinator can be allowed between the different local diagnosers. There are principally three notions to analyze the co-diagnosability property of systems: local, independent and decentralized diagnosability (see Fig. 4.2). These notions differ according to the interconnections between the system components. The local diagnosability property is the strongest property (see Fig. 4.2). If a system is local diagnosable then it is independent and decentralized diagnosable.

This book focused on the study of decentralized diagnosis approaches of discrete event systems. They are divided into two main categories: (1) the approaches using a global model in order to construct the local diagnosers and (2) the approaches using local models in order to construct the local diagnosers. Each of these two categories is divided into two main subcategories: the approaches considering a fault as the execution of an event and the approaches considering a fault as the violation of a specification as we can see in Fig. 4.3. Table 4.1 shows a comparison of these methods according to the system size, the co-diagnosability properties that can be verified and the kind of applications that they can use for.

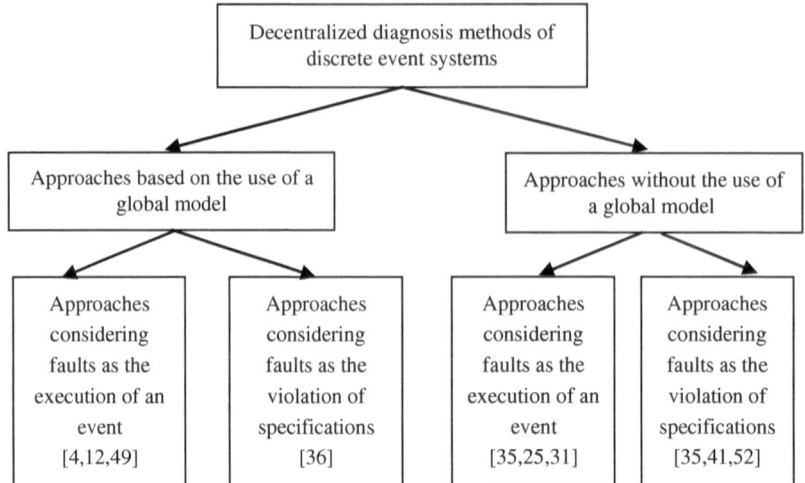

Fig. 4.3 Classification of the decentralized fault diagnosis approaches of discrete event systems

Table 4.1 Comparison of decentralized diagnosis approaches of discrete event systems

	Approaches considering faults as the execution of an event	Approaches considering faults as the violation of specifications
System size	Suitable whatever the size of the system	Suitable whatever the size of the system
Co-diagnosability property verification	Local, Independent and decentralized diagnosability	Local, Independent and decentralized diagnosability
Applications	Controlled and uncontrolled systems	Controlled systems
	Approaches based on the use of a global model	Approaches without the use of a global model
System size	Systems containing a small number of interconnected components or subsystems	Systems containing whatever the number of interconnected components or subsystems
Co-diagnosability property verification	Decentralized diagnosability	Local, Independent and decentralized diagnosability
Applications	Robotic, manufacturing systems, Production systems, etc.	Telecommunication networks, Power distribution networks, transport systems, etc.

References

1. Basile F, Chiacchio P, De Tommasi G (2009) An efficient approach for online diagnosis of discrete event systems. IEEE T Automat Contr 54(4):748–759
2. Basilio J-C, Lafortune S (2009) Robust codiagnosability of discrete event systems. American Control Conference, pp 2202–2209
3. Basilio J-C, Souza Lima S-T, Lafortune S, Moreira M-V (2012) Computation of minimal event bases that ensure diagnosability. Discrete Event Dyn Syst 22(3):249–292
4. Boel R-K, Van Schuppen J-H (2002) Decentralized failure diagnosis for discrete-event systems with costly communication between diagnosers. 6th International Workshop on Discrete Event Systems, pp 175–181
5. Cassandra C-G, Lafortune S (2008) Introduction to Discrete Event Systems, 2nd edn. Springer, New York Inc
6. Cabasino M-P, Giua A-N, Seatzu C (2010) Fault detection for discrete event systems using Petri nets with unobservable transitions. Automatica 46(9):1531–1539
7. Cassez F, Tripakis S (2008) Fault diagnosis with static and dynamic observers. Fundamenta Informaticae 88(4):497–540
8. Contant O, Lafortune S, Teneketzis D (2006) Diagnosability of discrete event systems with modular structure. Discrete Event Dyn Syst 16:9–37
9. Cordier M-O, Grastien A (2007) Exploiting independence in a decentralised and incremental approach of diagnosis. 20th International Joint Conference on Artificial Intelligence, pp 292–297
10. Cordier M-C, Le Guillou X, Robin S, Roze L, Vidal T (2007) Distributed chronicles for on-line diagnosis of web services. 18th International Worshop on Principles of Diagnosis, pp 37–44
11. Darkhovski B, Staroswiecki M (2003) Theoretic approach to decision in FDI. IEEE T Automat Contr 48(5):853–858
12. Debouk R, Lafortune S, Teneketzis D (2000) Coordinated decentralized protocols for failure diagnosis of discrete event systems. Discrete Event Dyn Syst 10(1–2):33–86
13. Devillez A, Sayed Mouchaweh M, Billaudel P (2004) A process monitoring module based on fuzzy logic and Pattern Recognition. Int J Approx Reason 37(1):43–70
14. Dotoli M, Fanti M, Mangini A (2009) Fault detection of DES by petri nets and integer linear programming. Automatica 45(11):2665–2672
15. Duda R-O, Hart P-E, Stork D-E (2001) Pattern classification, 2nd edn. Wiley, New York
16. Fabre E, Benveniste A, Haar S, Jard C (2005) Distributed monitoring of concurrent and asynchronous systems. Discrete Event Dyn Syst 15(1):33–84
17. Garcia H-E, Yoo T-S (2005) Model-based detection of routing events in discrete flow networks. Automatica 41(4):583–594
18. Genc S, Lafortune S (2003) Distributed diagnosis of discrete-event systems using Petri nets. International Conference on Application and Theory of Petri Nets, pp 316–336
19. Hernandez-Flores E, Lopez-Mellado E, Ramirez-Trevino A (2011) Diagnosticability analysis of partially observable deadlock-free Petri nets. 3rd International Workshop on Dependable Control of Discrete Systems, pp 176–181

M. Sayed-Mouchaweh, *Discrete Event Systems,* SpringerBriefs in Electrical and Computer Engineering, DOI 10.1007/978-1-4614-0031-8,
© Author 2014

20. Isermann R (2005) Model-based fault-detection and diagnosis: status and applications. Annu Rev Control 29:71–85
21. Jackson P (1998) Introduction to expert systems. Addison-Wesley Longman Publishing Co. Inc., Boston
22. Jéron T, Marchand H, Pinchinat S, Cordier M-O (2006) Supervision patterns in discrete event systems. 17th International Workshop on Principal of Diagnosis, pp 117–124
23. Jiang S, Kumar R (2004) Failure diagnosis of discrete event systems with linear-time temporal logic specifications. IEEE T Automat Contr 49(6):934–945
24. Jiroveanu G, Boel R-K (2006) A distributed approach for fault detection and diagnosis based on time Petri nets. Math Comput Simulat 70(5):287–313
25. Lamperti G, Zanella M (2008) On processing temporal observations in monitoring of discrete-event systems. Enterprise Inform Syst 3(3):135–146
26. Lin F (1994) Diagnosability of discrete event systems and its applications. Discrete Event Dynamic Systems. Kluwer Academic Publishers, Boston
27. Lunze J, Schroder J (2004) Sensor and actuator fault diagnosis of systems with discrete inputs and outputs. IEEE T Syst Man Cyb 34(3):1096–1107
28. Paoli A, Lafortune S (2005) Safe diagnosability for fault tolerant supervision of discrete-event systems. Automatica 41(8):1335–1347
29. Pandalai D, Holloway L-E (2000) Template languages for fault monitoring of timed discrete event processes. IEEE T Automat Contr 45(5):868–882
30. Patton R, Clark R-R, Frank M (2000) Issues of fault diagnosis for dynamic systems. Springer, Berlin
31. Pencolé Y (2000) Decentralized diagnoser approach: application to telecommunication networks. International Workshop on Principles of Diagnosis (DX'00), pp 185–192
32. Pencolé Y (2004) Diagnosability analysis of distributed discrete event systems. European Conference on Artificial Intelligence, pp 43–47
33. Pencolé Y, Cordier M-O (2005) A formal framework for the decentralised diagnosis of large scale discrete event systems and its application to telecommunication networks. Artif Intell 164(1–2):121–170
34. Perrow C (1984) Normal accidents: living with high risk technologies. Basic Books Inc., New York
35. Philippot A, Sayed Mouchaweh M, Carré Ménétrier V, Riera B (2011) Generation of candidates' tree for the fault diagnosis of discrete event systems. Control Eng Pract 19(9):1002–1013
36. Qiu W, Kumar R (2006) Decentralized failure diagnosis of discrete event systems. IEEE T Syst Man Cyb 36(2):628–643
37. Ramirez-Trevino A, Ruiz-Beltran E, Rivera-Rangel I, Lopez-Mellado E (2007) Online fault diagnosis of discrete event systems: a Petri net-based approach. IEEE T Autom Sci Eng 4(1): 31–39
38. Rozé L, Cordier M-O (2002) Diagnosing discrete-event systems: extending the "Diagnoser Approach" to deal with telecommunication networks. Discrete Event Dyn Syst 12(1):43–81
39. Sampath M, Sengupta R, Lafortune S, Sinnamohideen K, Teneketzis D (1995) Diagnosability of discrete event systems. IEEE T Automat Contr 40(9):1555–1575
40. Sampath M, Lafortune S, Teneketzis D (1998) Active diagnosis of discrete event systems. IEEE T Automat Contr 43(7):908–929
41. Sayed Mouchaweh M, Philippot A, Carre-Menetrier V (2008) Decentralized diagnosis by boolean discrete event system model: application on manufacturing systems. Taylor and Francis. Int J Prod Res 46(19):5469–5490
42. Sayed Mouchaweh M (2010) Semi supervised classification method for dynamic applications. Fuzzy Set Syst 161:544–563
43. Sayed Mouchaweh M (2012) Decentralized fault free model approach for fault detection and isolation of discrete event systems. Eur J Control 18(1):1–12
44. Sengupta R (1998) Diagnosis and communications in distributed systems. International Workshop on Discrete Event Systems, pp 144–151

45. Sengupta R, Tripakis S (2002) Decentralized diagnosability of regular languages is undecidable. 40th IEEE Conference on Decision and Control, pp 423–428
46. Su R, Wonham W-M (2005) Global and local consistencies in distributed fault diagnosis for discrete event systems. IEEE T Automat Contr 50(12):1923–1935
47. Takai S (2010) Robust failure diagnosis of partially observed discrete event systems. 10th International Workshop on Discrete Event Systems, pp 205–210.
48. Thorsley D, Teneketzis D (2007) Active acquisition of information for diagnosis and supervisory control of DES. Discrete Event Dyn Syst 17:531–583
49. Wang Y, Yoo T-S, Lafortune S (2007) Diagnosis of discrete event systems using decentralized architectures. Discrete Event Dyn Syst 17(2):233–263
50. Zad S-H, Kwong R-H, Wonham W-M (2003) Fault Diagnosis in Discrete Event Systems: framework and model reduction. IEEE T Automat Contr 48(7):1199–1212
51. Zad S-H, Kwong R-H, Wonham W-M (2005) Fault diagnosis in discrete-event systems: incorporating timing information. IEEE T Automat Contr 50(7):1010–1015
52. Zhou C, Kumar R, Sreenivas R-S (2008) Decentralized modular diagnosis of concurrent discrete event systems. 9th International Workshop on Discrete Event Systems, pp 388–393